Science for Sustainable Societies

Scope of the Series

This series aims to provide timely coverage of results of research conducted in accordance with the principles of sustainability science to address impediments to achieving sustainable societies – that is, societies that are low carbon emitters, that live in harmony with nature, and that promote the recycling and re-use of natural resources. Books in the series also address innovative means of advancing sustainability science itself in the development of both research and education models.

The overall goal of the series is to contribute to the development of sustainability science and to its promotion at research institutions worldwide, with a view to furthering knowledge and overcoming the limitations of traditional discipline-based research to address complex problems that afflict humanity and now seem intractable.

Books published in this series will be solicited from scholars working across academic disciplines to address challenges to sustainable development in all areas of human endeavors.

This is an official book series of the Integrated Research System for Sustainability Science (IR3S) of the University of Tokyo.

More information about this series at http://www.springer.com/series/11884

Kazuhiko Takeuchi • Hideaki Shiroyama
Osamu Saito • Masahiro Matsuura
Editors

Biofuels and Sustainability

Holistic Perspectives for Policy-making

Springer Open

Editors
Kazuhiko Takeuchi
Integrated Research System for Sustainability Science (IR3S)
Institutes for Advanced Study (UTIAS)
The University of Tokyo
Tokyo, Japan

Institute for Global Environmental Strategies (IGES)
Kanagawa, Japan

Osamu Saito
United Nations University Institute for the Advanced Study of Sustainability (UNU-IAS)
Tokyo, Japan

Hideaki Shiroyama
Graduate School of Law and Politics
The University of Tokyo
Tokyo, Japan

Masahiro Matsuura
Graduate School of Governance Studies
Meiji University
Tokyo, Japan

ISSN 2197-7348 ISSN 2197-7356 (electronic)
Science for Sustainable Societies
ISBN 978-4-431-54894-2 ISBN 978-4-431-54895-9 (eBook)
https://doi.org/10.1007/978-4-431-54895-9

Library of Congress Control Number: 2017963003

Printed on acid-free paper

This Springer imprint is published by the registered company Springer Japan KK part of Springer Nature.
The registered company address is: Shiroyama Trust Tower, 4-3-1 Toranomon, Minato-ku, Tokyo 105-6005, Japan

Contents

Chapter 1
Introduction

Hirotaka Matsuda and Kazuhiko Takeuchi

1.1 Background and Character of Biofuel Production Expansion

Currently, the development of biofuel expansion is found worldwide. Any energy released from biomass through a chemical reaction is called bioenergy (Yamajji et al. 2000). "Biomass" does not only mean biotic mass or biotic standing stock in ecological science but also means biotic mass as an energy source because it has been considered an alternative energy of fossil fuel since the "oil shocks" in the early 1970s. There is no strict definition, but the generic term covers an accumulation of animal and plant resources, as well as waste materials from them, except fossil resources, from the view of energy resources (The Japan Institute of Energy 2009). Biofuels can produce bioenergy, but it is often thought to be a fuel for transportation and is in competition with food crops. The current biofuels for transportation are mainly bioethanol and bio-diesel. These are called first-generation biofuels. Most first-generation biofuels are produced through glycosylating, fermenting and distilling starch ingredients of maize, wheat and potato or through fermenting and distilling carbohydrate ingredients of sugarcane and beet. They are also produced

---

H. Matsuda (✉)
Graduate Program in Sustainability Science – Global Leadership Initiative (GPSS-GLI), Graduate School of Frontier Sciences/Integrated Research System for Sustainability Science (IR3S), Institutes for Advanced Study (UTIAS), The University of Tokyo, 7-3-1, Hongo, Bunkyo-ku, Tokyo 113-8654, Japan
e-mail: matsuda@k.u-tokyo.ac.jp

K. Takeuchi
Integrated Research System for Sustainability Science (IR3S), Institutes for Advanced Study (UTIAS), The University of Tokyo, Tokyo, Japan

Institute for Global Environmental Strategies (IGES), Kanagawa, Japan

K. Takeuchi et al. (eds.), *Biofuels and Sustainability*, Science for Sustainable Societies, https://doi.org/10.1007/978-4-431-54895-9_1

from poaceous feed crops such as paddy and sorghums (The Japan Institute of Energy 2009 and Ohijiri 2004). Currently used first-generation bio-diesel is produced from animal oil and fat such as beef fat and lard, as well as vegetable oil such as *Elaeis guineensis* (for palm oil), crucifer (for canola oil), soybean (soybean oil) and sunflower (sunflower oil) (The Japan Institute of Energy 2009 and Matsumura 2006).

One of the reasons for introducing biofuels worldwide is its features.

Carbon Neutral The Kyoto protocol treats biofuels as carbon neutral because CO_2 emission for biofuel combustion is balanced out by absorbed CO_2 during growth of the plants for biofuel. In other words, while CO_2, one of the global greenhouse gasses (GHG), is emitted by fossil fuel burning, the Kyoto protocol views CO_2 emission as absent when biofuel is burned. We hypothesize that utilizing biofuels is a countermeasure against global warming. The most important reason for introducing biofuels is this carbon-neutral character in principle.

Renewable Energy According to the definition of renewable energy by the National Renewable Energy Laboratory, renewable energy is a non-exhaustible resource similar to wind and solar power (National Renewable Energy Laboratory 2008). Biofuel is thought to be a renewable energy because it is from plants and is not exhausted unless retarding the growth of the plants.

Prevention of Air Pollution An incomplete combustion of gasoline is inhibited by the addition of ethanol, which contains oxygen. Oxygenated gasoline by adding biofuels or bioethanol into gasoline reduces carbon monoxide emission. According to the IEA (International Energy Agency), not only carbon monoxide emission but also carbon hydride and particulate matter are reduced by adding bioethanol to gasoline. The US EPA (Environmental Protection Agency) reports that emissions of carbon monoxide, carbon hydride and particulate matter are reduced, although nitrogen oxide emissions are reduced by adding bioethanol to gasoline.

Contribution to Energy Security The production of biofuel from plants grown in a country may contribute to energy security in that country. In addition, producing biofuel can reduce the geopolitical risk of energy because of its even distribution, whereas fossil fuel often has high risk.

Development of Agriculture and Rural Areas Increasing agricultural profit, generating job opportunities in the agricultural sector and exporting agricultural production in developing countries are expected by biofuel production (Koizumi 2007; Hisano 2008).

These factors induce many countries and regions to focus on introducing biofuel production. In addition, a rapid increase in crude oil prices is also one of the major reasons for introducing biofuel production in many nations and regions. A decrease in the relative price of biofuel compared to the crude oil price as a substitute good leads to an increase in demand for it.

1.2 Current Situation of Biofuel Production in the World

According to the OECD-FAO (2013), the bioethanol production in the USA in 2013 is 55,769.8 million litres, and in Brazil, it is 28,684.5 million litres. The total production of both occupies 74.2% (USA, 48.9% and Brazil, 25.2%) of world production, 113,853.8 million litres. This trend is similar for the productions and shares of both countries; those of the USA are 79,997.3 million litres, 47.8%, and those of Brazil are 47,375.9 million litres, 28.3%, in 2022. Biofuel production has a long history. For instance, bioethanol was used for the Ford Model T in 1919, and blending bioethanol into gas was made obligatory in Brazil in 1931. However, increasing biofuel production in many countries and regions except Brazil is currently a possibility. The Energy Policy Act of 2005 and the Renewable Fuel Standard of the USA by President George W. Bush in 2005 as a midterm policy direction of energy in the USA and the State of the Union address by President Bush in 2006 and 2007 have had large impacts on biofuel policy in many countries and regions.

Bio-diesel productions of countries and regions in 2013 are as follows: EU27, 11,287.6 million litres (39.6%); USA, 6057.5 million litres (21.2%); Brazil,2405.0 million litres (11.5%); and Argentina, 2697.1 million litres (9.5%). This means that bio-diesel production is concentrated in a few countries and regions. Although it is expected that India produces a rather large amount of bio-diesels, it only produces 776.3 million litres (1.9%). EU27 is expected to produce 18,281.6 million litres (45.0%) along with the USA at 6267.2 million litres (15.4%), Brazil at 3336.6 million litres (8.2%) and Argentina at 3451.4 million litres (8.2%) in 2022. Bio-diesel production has a long history, as is the case of bioethanol. Although small-scale bio-diesel production was produced and used from the 1930s in some parts of the world, rapidly increasing bio-diesel production has been seen since approximately 2005, as is the case of bio-diesel.

While biofuel production has increased all over the world based on the futures indicated in Sect. 1.1, there is scepticism of the features. Promoting biofuel production may not only increase food supply and demand with adverse effects on agricultural production but also accelerate global warming.

1.3 Issues of Biofuels

The Kyoto protocol treats biofuels as carbon neutral; however, the whole producing process of biofuels, what we call the life cycle, should be evaluated. This process includes the energy input of agricultural production and energy crops for biofuels. Hill et al. (2006) estimated the energy balance of bioethanol production with DDGS (Distiller's Dried Grains with Solubles) from maize in 11 input cases. In addition,

Hill et al. compared these results with five existing papers. Although it is difficult to compare directly because inputs and products are different among studies, producing excess input energy was shown in four of six studies, Wang et al., Shapouri et al. (2004), Graboski (2002) and Hill et al. (2006). On the other hand, two of the six studies, Parikka (2004) and Pimentel (2003), have opposite results. Based on those studies, a clear result has not been obtained in terms of carbon-neutral biofuel production from the viewpoint of the life cycle. Hill et al. noted that those results are not derived from a common consensus of included inputs for biofuel production. For instance, it is difficult to define the ratio of agricultural capital use for biofuel crops from total inputs of agricultural capital for agricultural production. The UN-GBEP (Global Bioenergy Partnership) and many other institutions, however, have discussed a unified evaluation method of biofuel production that may be established. It is expected to establish international standards to evaluate biofuel production (Technical Innovation Council on Biofuels 2008).

Not only the energy balance of biofuel production but also the greenhouse gas emissions from soil are important in producing energy crops in the field. With greenhouse gas emissions from the cultivation of energy crops, the affirmation of carbon-neutral bioenergy may not be held.

Expanding the demand for bioenergy provides an incentive for farmers to shift current crop production systems to new crop production systems with energy crops. In fact, the number of farmers who do not sign up for the CRP (Conservation Reserve Program) in the United States is currently increasing. The CRP was started in 1986 to shift agricultural land located in disadvantaged areas to grass fields or forests. Some of the benefits from the CRP are increasing stored carbon in the soil, maintaining the productivity of land, mitigating land degradation caused by water and wind and protecting biodiversity. Extensional expansion of energy crops may drain benefits from the CRP. As a result, reducing greenhouse emissions through using biofuels, which is the most important projected contribution, is not only expected but also adversely affected by agricultural production through decreasing productivity of the land and the loss of biodiversity. In addition, it is noted that increasing agricultural production based on economic incentives leads to excess inputs of chemical fertilizer and pesticides (Fike et al. 2006; Parrish and Fike 2005). The increasing pricing pressure caused by the increasing demand for biofuels likely brings the same consequences. Increasing energy crop production with excess inputs could lead to harmful effects for ecological systems, including water systems.

It is expected that the so-called second-generation biofuels may alleviate the tight food supply because of biofuel expansion. Second-generation biofuels are produced from lignocellulosic biomass. Lignocellulosic biomass is hemicellulose, lignin and lignifying tissue, which are cells in the blade and stem (McKendry 2002). Although it takes time to put them into practical use, second-generation biofuels are expected to avert acute competition between crops for food and crops for biofuels since any part of crops except the edible part and agricultural residue may be used to produce biofuels. Additionally, second-generation biofuels are projected to produce larger amounts of biofuels than current biofuel production because larger parts of crops might be converted into biofuels in the case of second-generation biofuels than in the case of current biofuels (Perlack et al. 2005; Sheehan et al. 2004). While waiting for the introduction and dissemination of second generation of biofuels, increasing energy crop production might be prospected even by introducing second-generation biofuels. This means that increasing crop production based on economic incentives may not avoid greenhouse gas emissions from land or decreased land productivity and environmental deteriorations by excess inputs of chemical fertilizers and pesticides. It should also be noted that converting any part of the crops other than edible parts into biofuels might not maintain land productivity and carbon sequestration in the soil since turning the residues of crops such as maize, wheat and paddy into soil may contribute to maintaining that sequestration.

1.4 Biofuels and Sustainability Science

As discussed, biofuel utilization has a complex background and has broad impacts on many fields and sectors, such as the environment, economics and society. Therefore, a sustainable biofuel development strategy that may contribute to sustainable society is possible only if established by analysing the complex features of biofuels in a comprehensive manner.

The concept of sustainability has been discussed since sustainable development was discussed in the WCED (World Commission on Environment and Development) in 1987, which is known as the Brundtland Commission led by the Prime Minister of Norway, Brundtland (Maeda and Hibiki 2008). Through active debate in international arenas such as the UNCED (United Nations Conference on Environment and Development) and WBCSD (World Business Council for Sustainable Development), the atmosphere of building sustainability science, which is required to maintain a fundamental link between science and technology without policy bias, has been globally enhanced in academia (Komiyama and Takeuchi 2006). These active debates for sustainability science developed a common recognition of the need for transboundary/transdisciplinary academic systems that are different from traditional academic systems segmentalized in each academic field. A definition of sustainability science is propounded by Kates et al. based on historical debate and common recognition. The definition of sustainability science is that sustainability science sets out to solve global agendas of human subsistence such as global warming from the perspective point of sustainability (Maeda and Hibiki 2008).

A feature of sustainability science is solution-oriented science. Therefore, various research results and various researchers from many academic fields are joined in a transboundary/transdisciplinary way to solve global agendas. Global warming, for instance, is a problem shared by the entire human race that cannot be resolved by existing traditional approaches on a disaggregated basis, such as independent analysis regarding individual issues in individual regions and partial optimization analysis.

Sustainability science is still on the way to be mature in Europe, the United States and Japan. However, a common feature of sustainability science in academia is that the transboundary/transdisciplinary approach should be applied to resolve the issues that have multitiered and complex features by taking hold of those relationships (Komiyama and Takeuchi 2006; Clark and Dickson 2003; Kates et al. 2001; Lele 1991). In addition, resolving global agendas by applying sustainability science includes coordinating the related stakeholders.

While research results for the effects of biofuels on the environment from the natural science view have accumulated gradually, there is still room for biofuels research to be analysed. It should be considered to consolidate not only existing scientific results regarding biofuels but also new scientific knowledge to policymakers and stakeholders as scientific evidence. Biofuel utilization should be considered a trilemma of global warming, energy security and food security, promoting agriculture in other words. Moreover, biofuel utilization is seen as one of the factors of acute food price increases. It is imperative to coordinate among international institutions, policymakers across nations and other stakeholders to establish a sustainable biofuel development strategy based on an adaptation/mitigation strategy from various scientific knowledge for biofuel utilization. Applying the concept of sustainability science allows us to build that strategy. Meanwhile, applying sustainability science to establish a sustainable biofuel development strategy may contribute to an increasing global stream of building sustainability science.

1.5 Objectives

As discussed, biofuel utilization has a complex background and has broad impacts on many fields and sectors, such as the environment, economics and society. Therefore, a sustainable biofuel development strategy that may contribute to sustainable society is only possible by analysing the complex features of biofuels in a comprehensive manner. It is necessary to integrate the findings from the analysis of social sciences and natural sciences.

The objectives of this book develop a development strategy for biofuels at the multi-scale, national, regional and worldwide levels through integrating analysis by social sciences and natural sciences based on a sustainability science approach. As mentioned in other chapters, the feature of sustainability science is that various research results and various researchers from many academic fields are joined in a transboundary/transdisciplinary way to solve global agendas. Therefore, sustainability science is better suited for analysing biofuels that have a wide-ranging impact and establishing a sustainable development strategy.

To achieve our aims, this book has three main parts. In part I, the conceptual framework of this book is shown. Research results for biofuels from the views of natural science and social science are indicated in part II. Research has been conducted at the multi-scale, global, regional and national levels. Our main focus is the Asia Pacific region, including China, India, Indonesia and Japan. In part III, sustainable biofuel development strategies at the multi-scale level are shown as a result.

References

Clark WC, Dickson NM (2003) Sustainability science: the emerging research program. PNAS 100(14):8059–8061

Fike JH, Parrish DJ, Wolf DD, Balasko JA, Green JT, Rasnake M, Reynolds JH (2006) Long-term yield potential of switchgrass-for-biofuel systems. Biomass Bioenergy 30(3):198–206

Graboski MS 2002 Fossil energy use in the manufacture of corn ethanol. Prepared for the National Corn Growers Association, St Louis, Missouri, USA.

Hill J, Nelson E, Tilman D, Polasky S, Tiffany D (2006) Environmental, economic, and energetic costs and benefits of biodiesel and ethanol biofuels. Proc Natl Acad Sci 103(30):11206–11210

Hisano S (2008) Political economy of biofuel boom –how clean is the green? Agric Agric Cooper Stud 38:16–27

Kates RW, Clark WC, Corell R, Hall JM, Jaeger CC, Lowe I, McCarthy JJ, Schellnhuber HJ, Bolin B, Dickson NM, Faucheux S, Gallopin GC, Grübler A, Huntley B, Jäger J, Jodha NS, Kasperson RE, Mabogunje A, Matson P, Mooney H, Moore B III, O'Riordan T (2001) Environment and development: sustainability science. Science 292(5517):641–642

Koizumi T (2007) Bioethanol and food demand-supply in the world, Tsukuba Shobo

Komiyama H, Takeuchi K (2006) Sustainability science: building a new discipline. Sustain Sci 1(1):1–6

Lele SM (1991) Sustainable development: a critical review. World Dev 19(6):607–621

Maeda S, Hibiki S (2008) Research trend of sustainability science for globalwarming. Sci Technol Trends 84. http://www.nistep.go.jp/achiev/ftx/jpn/stfc/stt084j/0803_03_featurearticles/0803fa01/200803_fa01.html

Matsumura M, Sun Care Fuels Corporation (eds) (2006) Graphic explanation frontier of biodiesel. Kogyo Chosakai Publishing Co., Ltd

McKendry P (2002) Energy production from biomass (part 1): overview of biomass. Bioresour Technol 83(1):37–43

National Renewable Energy Laboratory (2008) http://www.nrel.gov/

OECD-FAO (2013) OECD-FAO Agricultural outlook 2013–2022. http://www.oecd.org/site/oecd-faoagriculturaloutlook/

Ohijiri Y, Mitsui & CO., LTD (eds) (2004) Graphic explanation frontier of biodiesel. Kogyo Chosakai Publishing Co., Ltd..

Parikka M (2004) Global biomass fuel resources. Biomass Bioenergy 27:613–620

Parrish DJ, Fike JH (2005) The biology and agronomy of switchgrass for biofuels. Crit Rev Plant Sci 24(5–6):423–459

Perlack RD, Wright LL, Turhollow A, Graham RL, Stokes B, Erbach DC (2005) Biomass as feedstock for a bioenergy and bioproducts industry: the technical feasibility of a billion-ton annual supply. Oak Ridge National Laboratory, Oak Ridge, Tennessee, USA, DOE/GO-102995-2135, ORNL/TM-2005/66

Pimentel D (2003) Ethanol fuels: energy balance, economics, and environmental impacts are negative. Nat Resour Res 12(2):127–134

Shapouri H, Duffield J, McAloon A, Wang A (2004) The 2001 net energy balance of corn-ethanol. USDA, Washington, DC

Sheehan J, Paustian K, Walsh M, Nelson R (2004) Energy and environmental aspects of using corn stover for fuel ethanol?. J Ind Ecol 7(3–4):117–146
Technical Innovation Council on Biofuels (2008) Strategy of biofuel technical innovation
The Japan Institute of Energy (2009) Biomass handbook. Ohmsha, Tokyo
Yamajji K, Yamamoto H, Fujino J (2000) Bioenergy. Mion Publishing

Part I
Biofuels and Sustainability Conceptual Framework

Chapter 2
Approach to Biofuel Issues from the Perspective of Sustainability Science Studies

Hirotaka Matsuda and Kazuhiko Takeuchi

2.1 Introduction

Biofuels have been increasing in popularity, since they are promising substitutes for fossil fuels and are expected to contribute to reductions in greenhouse gas (GHG) emissions. Moreover, the production of biofuels is a means of alleviating poverty and developing both rural and agricultural areas. However, many researchers and institutions, such as the Organization for Economic Co-operation and Development (OCED) and the Food and Agriculture Organization (FAO), voice scientific scepticism about the expected contributions of biofuel use. They also stress that the production and use of biofuels will lead to deforestation, water supply contamination and water depletion. The production and use of biofuels will have enormous impacts on the environment, the economy and the society. Clearly, these impacts are multitiered and complex. Therefore, strategies for biofuel use must be established through comprehensive analyses and scientific evaluations, with consideration given to complex socioeconomic issues, in order to achieve global sustainability. It is also important to consider that optimum solutions among boundary levels, such as global, regional and national levels, may vary and that these strategies must be coordinated in order to meet the demands of different optimum solutions. From this perspective, an interdisciplinary and integrated approach is best. However, many

H. Matsuda (✉)
Graduate Program in Sustainability Science – Global Leadership Initiative (GPSS-GLI), Graduate School of Frontier Sciences/Integrated Research System for Sustainability Science (IR3S), Institutes for Advanced Study (UTIAS), The University of Tokyo,
7-3-1, Hongo, Bunkyo-ku, Tokyo 113-8654, Japan
e-mail: matsuda@k.u-tokyo.ac.jp

K. Takeuchi
Integrated Research System for Sustainability Science (IR3S), Institutes for Advanced Study (UTIAS), The University of Tokyo, Tokyo, Japan

Institute for Global Environmental Strategies (IGES), Kanagawa, Japan

K. Takeuchi et al. (eds.), *Biofuels and Sustainability*, Science for Sustainable Societies, https://doi.org/10.1007/978-4-431-54895-9_2

studies on biofuel, including those in the natural and social science fields, fail to use this type of approach. The aim of the present research is to comprehensively analyse the use of biofuels at global, regional and national levels using the sustainability science approach and attempt to assess biofuel use strategies from an interdisciplinary perspective. Sustainability science is a new academic area that addresses complicated issues, such as biofuel production and use, by restructuring problems and then proposing policy options.

2.2 What Is the Sustainability Science?

As discussed, biofuel utilization has a complex background and broad impacts on many fields and sectors, such as the environment, the economy and the society. Therefore, the establishment of a sustainable biofuel strategy that contributes to a sustainable society is only possible by analysing the complex features of biofuels in a comprehensive manner.

The concept of sustainability has previously been discussed, since sustainable development was discussed in the WCED (World Commission on Environment and Development) in 1987, an event known as the Brundtland Commission that was led by the Prime Minister of Norway, Brundtland (Maeda and Hibiki 2008). Through active debate in an international arena such as the UNCED (United Nations Conference on Environment and Development) and WBCSD (World Business Council for Sustainable Development), the atmosphere of building sustainability science required to maintain fundamental links between science and technology without policy bias has been enhanced in academia globally (Komiyama and Takeuchi 2006). These active debates for sustainability science have developed a common recognition of the need for transboundary/transdisciplinary academic systems, which are different from traditional academic systems that are segmentalized in each academic field. A definition of sustainability science was propounded by Kates et al. based on the historical debate and common recognition. This definition states that sustainability science sets out to solve global agendas of human subsistence, such as global warming, from the view point of sustainability (Maeda and Hibiki 2008).

A feature of sustainability science is solution-oriented science. Therefore, various research results are brought to various researchers from many academic fields in a transboundary/transdisciplinary manner to solve global agendas. For example, global warming, which is a problem shared by the entire human race, cannot be resolved by existing traditional approaches on a disaggregated basis, such as independent analyses regarding individual issues for individual regions or partial optimization analysis.

The development of sustainability science, which is being led by Europe, the United States and Japan, is still ongoing. However, a common feature of sustainability science in academia is that a transboundary/transdisciplinary approach should be applied to resolve issues that have multitiered and complex features by recognizing those relationships (Komiyama and Takeuchi 2006; Clark and Dickson 2003; Kates et al. 2001; Lele 1991). In addition, resolving global agendas by the application of sustainability science includes the coordination of related stakeholders.

Although the IPCC (Intergovernmental Panel on Climate Change) has an influence on the establishment of sustainability science, its role and existence are affected by the discussion of sustainability science. An extremely significant contribution of IPCC is its presentation of the impact of global warming as anthropogenic, which became a united opinion due to the research evidence that the IPCC amassed. That scientific knowledge has contributed to policy decision-making processes by nations and international institutions, including the UNFCCC (United Nations Framework Convention on Climate Change). Currently, the role of science has moved from the clarification of the global warming phenomenon to the building of adaptation and mitigation strategies for global warming.

Sachs and Reid note that an investment in poverty reduction is critical for environmental policy. Furthermore, they also note that an investment in the environment is important for the success of poverty alleviation. In addition, they insist that a global assessment scheme for mutual relationships between poverty alleviation and environment protection should be established by the United Nations, IPCC and MEA (Millennium Ecosystem Assessment). They advocate that a global network of scientists, including environmentalists, economists and social scientists, can inform policy makers and the general public of the latest scientific findings and that the network can additionally overcome the opaqueness originating from vested interest groups by structuring required research freely. Therefore, strategies built on trans-boundary/transdisciplinary foundations are needed for sustainable development. An affirmation of Sachs and Reid is believed to be the links among poverty alleviation, agricultural production, and sustainability science.

2.3 Feature of Biofuels from the Sustainability Science View

Biofuel features are reported in this section from the sustainability science viewpoint.

Biofuel impacts are spread across a wide area. First, an impact of biofuels on the economy is noted. Since 2006, "agflation" has become a serious problem all over the world. It is noted that biofuels are seen as one of the factors contributing to agflation. Although further research on the relationship between agflation and biofuels is required, it is undeniable that biofuels cause agflation. As a result, many developing countries are in socio-political dislocation. Some of these countries regulate food export and agricultural prices. Although those policies tend to be chosen from the view point of food security in these countries, agflation threatens to shrink the international cereal market and further increase pricing pressure. The poorest segments of the population experience difficulties obtaining food because of agflation. As the FAO notes in Food Outlook 2007 (FAO 2008), this situation leads to further socio-political confusion in LDC (least developed countries), LIFDC (low-income food-deficit countries) and NFIDC (net food-importing developing countries). However, a rise in the price of agriculture may stimulate agricultural production in both developing countries and developed countries.

It is noted that the extensional expansion of agricultural production for biofuels might not only fail to contribute to reductions in GHG emissions because of the outflow of carbon storage in the soil but also have adverse effects on agricultural production because of biodiversity loss and decreased land productivity. Furthermore, increasing agricultural production on the basis of economic incentives induces the use of chemical fertilizers and pesticides (Fike et al. 2006; Parrish and Fike 2005). Increases in agricultural production resulting from economic incentives seem to be predominant, which is inferred to induce adverse effects on the ecosystem.

The consideration of importing biofuels and agricultural products for biofuels by Japan, EU and some other countries is subjected to criticism, since the import of biofuels and agricultural products for biofuels that are derived from agricultural production in developing countries promotes environmental degradation. A valid judgement is required for this issue. However, it cannot be denied that increased agricultural production for exports plays a role in rural development. Areas with high levels of environmental degradation have an advantage for biofuel production. Biofuel production or agricultural production for biofuels in those areas might improve the welfare of the world in terms of the efficiency of resource allocation (FAO 2008).

2.4 Conclusion

While research results on the effects of biofuels on the environment from the natural science perspective have accumulated gradually, there is still room for biofuel research to be analysed. Not only existing scientific results regarding biofuels but also new scientific knowledge should be consolidated for policymakers and stakeholders as scientific evidence. Biofuel utilization should be considered a trilemma of global warming, energy security and food security, the promotion of agriculture, in other words. Moreover, biofuel utilization is seen as one of the factors contributing to an acute increase in food prices. It is imperative to coordinate among international institutions, policymakers in many nations and other stakeholders to establish sustainable biofuel utilization strategies based on adaptation/mitigation strategies supported by various scientific results on biofuel utilization. Applying the concept of sustainability science allows us to build these strategies. In addition, applying sustainability science to establish sustainable biofuel utilization strategies may contribute to the global increase in building sustainability science.

References

Clark WC, Dickson NM (2003) Sustainability science: the emerging research program. PNAS 100(14):8059–8061

FAO (2008) The state of food and agriculture 2008. Biofuels: prospects, risks and opportunities. FAO, Rome

Fike JH, Parrish DJ, Wolf DD, Balasko JA, Green JT, Rasnake M, Reynolds JH (2006) Long-term yield potential of switchgrass-for-biofuel systems. Biomass Bioenergy 30(3):198–206

Kates RW, Clark WC, Corell R, Hall JM, Jaeger CC, Lowe I, McCarthy JJ, Schellnhuber HJ, Bolin B, Dickson NM, Faucheux S, Gallopin GC, Grübler A, Huntley B, Jäger J, Jodha NS, Kasperson RE, Mabogunje A, Matson P, Mooney H, Moore B III, O'Riordan T (2001) Environment and development: sustainability science. Science 292(5517):641–642

Komiyama H, Takeuchi K (2006) Sustainability science: building a new discipline. Sustain Sci 1(1):1–6

Lele SM (1991) Sustainable development: a critical review. World Dev 19(6):607–621

Maeda S, Hibiki S (2008) Research trend of sustainability science for global warming. Sci Technol Trends 84. http://www.nistep.go.jp/achiev/ftx/jpn/stfc/stt084j/0803_03_featurearticles/0803fa01/200803_fa01.html

Parrish DJ, Fike JH (2005) The biology and agronomy of switchgrass for biofuels. Crit Rev Plant Sci 24(5–6):423–459

Chapter 3
Stakeholder Perspective and Multilevel Governance

Masahiro Matsuura and Hideaki Shiroyama

3.1 Stakeholder Perspectives

3.1.1 Defining Who the Stakeholders Are

In the field of public policy analysis, the concept of "stakeholders" has been widely applied to a variety of policy-making efforts. In particular, the stakeholder concept has been adopted in the shift of focus from the government to the governance. In this context, traditional bureaucratic government structure endowed with the power of "command and control" is regarded inefficient anymore in the democratic and internationalized environment. Networked actors that undertake the functions previously performed by the government would replace the traditional structure. In this new "governance"-focused system, stakeholders, instead of the government, undertake the public sector functions. In other words, stakeholders are the individuals and organizations that actively participate in policy-making processes and take appropriate responsibilities of implementing the policies that they have agreed to.

The definition of stakeholders, however, has not been discussed much in the field of public policy. The same concept is often represented by other terms such as "actors" and "players." In the field of corporate management, the definition of stakeholders was initially proposed R. E. Freeman, who is currently considered as the pioneer in the field of stakeholder-focused management. He argues that stakeholders are those who have influence in decision-making and those who are influenced

M. Matsuura (✉)
Graduate School of Governance Studies, Meiji University,
1-1 Kanda Surugadai, Chiyoda-ku, Tokyo 101-8301, Japan
e-mail: mmatsuura@meiji.ac.jp

H. Shiroyama
Graduate School of Law and Politics, The University of Tokyo, Tokyo, Japan

K. Takeuchi et al. (eds.), *Biofuels and Sustainability*, Science for Sustainable Societies, https://doi.org/10.1007/978-4-431-54895-9_3

by the decisions (Freeman 1984). The broad definition of stakeholders suggests the importance of having a holistic picture of a wide range of in the decision-making environment that might appear to be dominated by a few executives. Freeman regards stakeholder management as an opportunity for value creation through developing collaborative relationships with stakeholders external to the organization in focus.

The same principle can be applied to varieties of studies in the field of public policy. The term stakeholder encompasses a wide range of organizations and individuals that have, either direct or indirect, relationships with the policy and decision that policy analyst is concerned about. It should not be limited to the formal organizations that have statutory rights to participate and/or veto. Albeit this narrow conception might be useful in legal studies, the boundary between who have the stake or not is quite obscure in the realm of politics. Therefore, any policy analysis with focus on stakeholders, for instance, should involve those organizations and individuals that have implication with the policy even if they have no formal rights to redress.

In the context of public policy, analyzing stakeholders has been particularly important at the relatively local level. For decisions pertaining to specific development projects, categories of stakeholders are often represented by specific organizations, corporations, and individuals. Case studies, as well as pragmatic analysis for convening stakeholder dialogues, identify these stakeholders and analyze the interaction between these specific stakeholders in policy-making processes. The category of stakeholders becomes less specific when the analysis of stakeholder is applied to national and international strategies. In such instances, a manageable number of broad categories of stakeholders are defined.

3.1.2 Applying the Stakeholder Perspective to the Biofuel Cases

When we apply this stakeholder perspective to analyzing the sustainable deployment of biofuels, the way of defining stakeholders can vary significantly. For example, if one intends to limit the focus to the distillation processes of sugarcane-based ethanol on Miyakojima Island in Japan, stakeholder categories would be represented by specific organizations or even individuals such as councilpersons and village heads. On the other hand, if we broaden the focus to the global strategy for the sustainable use of biofuels, including a wide range of feedstocks, as we intend in this book, stakeholder categories would be defined by the broad functions of stakeholders in the series of biofuel production and delivery processes. In order to limit the number of stakeholder categories at a practical level, organizations and individuals have to be bundled together under a certain category.

Stakeholder dialogues have already been convened in the context of sustainable deployment of biofuels. For instance, the Roundtable on Sustainable Biofuels (RSB), which is convened by the Energy Center at École Polytechnique Fédérale de

Lausanne, organizes seven chambers which correspond to their conception of stakeholders. They are (1) farmers and growers of biofuel feedstocks; (2) industrial biofuel producers; (3) retailers/blenders, transportation industry, and banks/investors; (4) rights-based NGOs (including land, water, human, and labor rights) and trade unions; (5) rural development or food security organizations and smallholder farmer organizations or indigenous peoples' organizations or community-based civil society organizations; (6) environment or conservation organizations and climate change or policy organizations; and (7) intergovernmental organizations (IGOs), governments, standard setters, specialist advisory agencies, certification agencies, and consultant experts. Under these headings, stakeholders from around the world convene to the roundtable and take responsibilities in developing and maintaining a global governance structure on the sustainable biofuels. A similar effort, Roundtable for Sustainable Palm Oil, defines stakeholders as "An individual or group with a legitimate and/or demonstrable interest in, or who is directly affected by, the activities of an organisation and the consequences of those activities," and encourages their participation through various consultation mechanisms (RSPO 2006).

3.1.3 *Stakeholder Perspective as an Essential Element of Good Policy Processes*

As the nations mature economically, the size of resources available to the government, in relation to the scale of national economy, shrinks. On the other hand, certain public services must be provided in order to maintain the nation as an association of free individuals. In this environment, public services, which were provided solely by the government sector, need to be restructured around a voluntary agreement among stakeholders including private corporations as well as civil society organizations. This trend has been particularly evident in Japan in the last few years. The current Democratic Party administration has been promoting "the new public (*atarashii ko-kyo*)" initiatives which attempt to minimize the direct involvement of the government – which has been pursed under the previous administration that can be characterized as the most neoliberal regime in the history of modern Japan – while addressing the public service needs through voluntary or civil society organizations. Rather than just letting the market decide, the new initiatives try to take care of the necessary public functions by fostering collaborations among the government, civil society organizations, as well as private corporations.

The same kind of collaborative arrangement is important in the realm of international *governance* because fundamentally all decisions are in reality based on voluntary agreements among nation-states and other stakeholding parties. Because of the Westphalian sovereignty of nation-states, no institution can force a nation to take a certain course of actions unless in extraordinary situations. Under this constraint, stakeholding parties in the global context need to reach a voluntary agreement that they can live with.

Therefore, under the systems of *governance*, policies and strategies can be conceptualized as a kind of voluntary agreements among stakeholders. In other words, any system of *governance* cannot guarantee its stable operation without consent by overwhelming number of stakeholders. This kind of voluntary arrangement, of course, is at the risk of collective action problems. Therefore, any stakeholder agreement must be accompanied by well-articulated mechanisms that prevent free riders from the framework.

Why do they have to reach an agreement, assuming that these stakeholders might be able to live alone without interacting with other stakeholders? Two kinds of argument are forthcoming. First, the mutual dependence between these stakeholders is so important in this global economy that an option of not collaborating with other stakeholders entails a massive loss or a huge risk. In particular, the volume of international trade has increased – for instance, by as much as 9.5% only in 1 year of 2010 – and every individual on the planet would be affected somehow by international agreements. For instance, how is it likely for a palm oil plantation owner to negate an internationally accepted sustainability standards on its production? Such a plantation owner can be easily expelled from the international market and will lose his/her competitiveness particularly because the crude palm oil is now one of the major internationally traded commodities. Not participating in world trade organization and other international mechanisms would risk the economy of a nation.

Climate change and other transboundary environmental issues are another representation of mutual dependence that brings nations together. Due to their massive size of externality, a variety of stakeholders need to make a commitment to a governance mechanism that circumvents the risk of catastrophes at the global scale. We, including the future generations, share a risk of so-called lose-lose outcome in the classic prisoner's dilemma situation.

Second, stakeholder collaboration can also be conceptualized as an opportunity for value creation. For instance, the involvement of nongovernmental organizations (NGOs) around the world in the implementation of global arrangement can reduce the cost of implementation and monitoring, compared to a supranational organization taking over the whole responsibility of implementation. This kind of networked governance can be sustained through the mutual gains to all parties involved in such arrangement.

Negotiated agreements are said to produce fair, efficient, stable, and wise solution, compared to the conventional command and control decisions (Susskind and Cruikshank 1987). One example is the negotiated rulemaking programs by the US Environmental Protection Agency. When the agency intends to issue a regulation, stakeholder representatives are convened to reach an agreement on a draft regulation. When the EPA issues the regulation by adopting the draft prepared by stakeholders, the risk of the EPA being sued for the regulation is lower because the stakeholders previously agreed to the regulation. Therefore, stakeholder-based approaches are far better than the traditional command and control approaches based on the rational.

3.1.4 Broader Conception of Stakeholders

In practice, however, the stakeholder perspective could be harmful for the evolution of democratic society. If one employs a narrow definition of stakeholders and limit the political participation to those who actually have the power to influence the decision or the access to redress, those who might be influenced by the decision but have no formal right to appeal are likely to be excluded simply because of the arbitrarily defined boundary of legitimate stakeholders.

For instance, future generations might not be considered as a legitimate category of stakeholders, leading to unrecoverable environmental damages. Indigenous people without political influence would be neglected as marginal actors. Such narrow conceptions of stakeholders might lead to a solution that strengthens the incumbent power structure that might not be "democratic" or "sustainable" at all.

Thus, the stakeholder perspective, if it is misconstrued, can be employed as a tool for the incumbents to amass their political influence. Meanwhile, those poor people who have no access to the political arena would have less access to policy-making processes where they could voice their concerns. Such concerns have led to the criticism about the conventional liberal conception of bargaining-based approaches to policy-making.

We, however, take a different approach. We assert that stakeholders should be conceptualized in a long-term and global perspective. Any strategy that merits the current generation and demand insurmountable burden on the future generation is not sustainable at all, as the Brundtland Commission concluded in its statement on sustainable development. Indigenous people deprived of political access under the current regime might gain political power with help of international actors, such as international nongovernmental organizations (INGOs), in a long run. Citizen's revolutionary movements, as we saw in some of the northern African countries in 2011, can lead to a dramatic change of domestic power structure.

In this regard, a concept called "activist mediator" is instructive. Conventionally, mediators try to resolve conflicts between specific parties under certain conditions. Forester and Stitzel (1989) argue, however, mediators in the public sector dispute resolution efforts take more proactive roles in resolving conflict. For instance, they try to involve stakeholders who are not necessarily identified as the main parties to the dispute. They also try to encourage the disputants to consider "other" stakeholders, such as future generations, so that their agreement can be sustainably implemented in the long run.

We take an activist mediator's approach to the stakeholder perspective. We argue that the conception of stakeholders should not be bounded by the current power structure that surrounds the policy situation of concern. Instead, anyone who tries to identify the range of stakeholders should imagine how the structural constraints, which define the range of stakeholders, might change in a long run. He/she should also give up being totally objective in the analysis and take a stand in involving those who should, instead of who can, participate in a democratic decision-making.

3.1.5 Why This Perspective Is Important in the Study of Biofuel Deployment

Involving a wide range of stakeholders contributes to an increased political stability of the strategy that we propose in this book. Any strategy that ignores the views of certain categories of stakeholder has the risk of having it overthrown sometime later due to their amounting discontent.

Stakeholder involvement can contribute to environmental justice. Particularly in developing nations, economic interests of the dominant parties can overshadow the voice of poor people. If we take the shortsighted neoliberal approach to dealing with the issue, their interests cannot be incorporated into our analysis because they do not have sufficient influence in the policy-making processes. However, if we take a long-term perspective for sustainable deployment of biofuels, it is necessary to recognize the opportunity for developing sustainable and democratic governance in these nations. Governance structure might shift over the time. In order to achieve a robust strategy, it is necessary to have a long-term stakeholder perspective.

Therefore, advocates of stakeholder perspective need to admit that such approach has an effect of empowering certain categories of stakeholders who are currently underrepresented. They should also bring other kinds of underrepresented stakeholders to the arena of deliberation.

Under the high level of uncertainty, our strategy should also be designed as an adaptive system that allows flexible rearrangements to the changing environment. In order to achieve that, stakeholders should also be continuously redefined, and their search for common ground should be embedded in a perpetual institution.

3.2 Multilevel Governance

3.2.1 Levels of Governance

Biofuel deployment requires a holistic analysis of stakeholders at different levels of governance. For instance, each consumer makes a choice between biofuel and conventional fossil fuel at the gas station. This action occurs at the very local level involving a number of consumers and gas stations. Meanwhile, imports of biofuels occur at the international level. While it might involve a limited number of stakeholders and transactions, it can have major impacts on the utilization of biofuels at the national and local levels. Therefore, it is necessary to look at biofuel utilization policy at different levels of governance, from the global to the local.

It is also necessary to look at the regional/national level as an intermediary between the global and local levels. At this level, for instance, public policy instruments of each country have influence on the utilization of biofuels. While biofuel has become a worldwide issue because of its implication on the global environment, still each national government has significant power in determining the course of its usage. Government agencies set the mandates, regulations, and other subsidies for biofuel usage in their countries. Such policies are debated by different stakeholders in each country, including civil society organizations, consumer groups, members of the petroleum industry, automobile producers, as well as local representatives of INGOs. Therefore, it is still necessary to look at individual regions and nations as a kind of boundary that sets the arena for biofuel policy-making.

3.2.2 Multilayered and Nested Nature of Biofuel Governance

Because the *governance* concept is grounded primarily on voluntary agreements between stakeholders, it can be identified at any level. International organizations and national representatives are key players in the governance at the global level. Individual consumers, gas station operators, and even manual laborers are the key stakeholders at the local level. At each of these levels, there have to be certain agreements among these stakeholders for these governance systems to sustain.

Thus, biofuel governance can be identified in a nested system of a multilayered environment. While each system of governance has to be grounded on a kind of social contract among stakeholders, individual systems of governance influence each other, and the coordination among them is another key factor in considering the sustainability of holistic systems for the utilization of biofuels. It is insufficient for a researcher to look at only one level of governance without studying its influence to the other levels as well as the influences that it might incur from the other levels.

Multilevel governance is an idea adopted particularly in the study of EU governance. The interaction between the EU and participating nation become the subject of research after its harmonization efforts started in the 1990s. Each member state has an obligation to follow the directives and decisions by the directorate general of the European Commission and the European Parliament. The direction of the influence is, however, not one-way. Each member state, as well as lobbyists sent by industries of each nation, tries to influence the EC policy in Brussels and Strasbourg. Thus, the influence is bidirectional. This interaction between nation-states and international organization has attracted the interests of European political scientists.

The same concept can be applied to the multilevel governance of biofuels. As we stated, it is a matter of policy and market decisions at the international, regional/national, and local levels. The interaction among governance systems at these three levels represents a complex tension among stakeholders at multiple levels.

3.2.3 Why This Perspective Is Important in the Study of Biofuel Deployment

Our strategy is robust because it reflects the realities of biofuel deployment at all levels. International arrangements need to be supported by enormous number of stakeholders in the field. Efforts at the local level must be supported and diffused nationally and internationally in order to have a large-scale impact. Multilevel governance perspective leads us to pay more attention to the interactions between different layers so that efforts at different levels can have a synergy effect.

Chapter 4
Applying Stakeholder Perspectives to Sustainable Biofuel Strategy: A Summary of Our Analyses

Masahiro Matsuura and Hideaki Shiroyama

4.1 Producers in Developing Nations

Toward the mass production of biofuels for transportation and other uses, feedstock production is increasingly dependent on developing nations in South America and Southeast Asia. For instance, multiple sections in Part II focused on the production of sugarcane-based bioethanol in Brazil. Chapter 2.2.1 will analyze the impact of increased production of sugarcane in Brazil on forest, land, and water uses. In a similar vein, Chap. 2.1.2 will discuss various methods of bioethanol production that would eventually contribute to the ultimate goal of deploying biofuels, which is to reduce the GHG emission. Chapter 2.2.2 will also discuss various methods of production with focus on regional impacts. Chapter 2.3.1 will provide an overview of stakeholders in Brazilian bioethanol and Indonesian biodiesel production sectors. These chapters focus on producers' influence on the environment, as well as the influence on varieties of stakeholders in the production of biofuels.

In the context of regulating biofuels, "producers" of feedstock are often characterized as profit-seeking plantation owners that contribute to the degradation of natural environment and living environment of indigenous people. The reality in the field of production in developing nations, however, is far more complex. Different kinds of plantation owners exist, varying by the scale of capital and the main market. Plantation owners are not the sole decision-maker in the feedstock production. Many independent small-scale farmers still exist.

M. Matsuura (✉)
Graduate School of Governance Studies, Meiji University,
1-1 Kanda Surugadai, Chiyoda-ku, Tokyo 101-8301, Japan
e-mail: mmatsuura@meiji.ac.jp

H. Shiroyama
Graduate School of Law and Politics, The University of Tokyo, Tokyo, Japan

K. Takeuchi et al. (eds.), *Biofuels and Sustainability*, Science for Sustainable Societies, https://doi.org/10.1007/978-4-431-54895-9_4

In addition, distilling and refining feedstock into biofuel is a major question in terms of profit, particularly in the case of Indonesian biodiesel. The refinery part of the biofuel production is far more profitable than feedstock production, while the former requires capital investments and technology. Therefore, some part of Indonesian crude palm oil is transferred to Singapore for final processes, which makes Indonesian stakeholders demand "fair" share. In the case of bioethanol in Brazil, biofuel production plants are often integrated with conventional sugar cane production plants. Therefore, research and development for better refinery system occurs in Brazil, which allows the country to fully benefit from increased production of bioethanol. Advanced technologies and their benefit to Brazilian communities will be further discussed in Chap. 2.1.2 (Table 4.1).

4.2 Users in Developing Nations

While discourses on biofuels are often focused on the increasing demand for biofuels at the global scale because of the need to offset GHG emissions, domestic users are in fact the major players in the deployment of biofuels. Brazil's Pró-Álcool policy in response to the oil crisis of 1973 was successful in achieving the market-scale production of bioethanols, supported by the introduction of flex-fuel technologies in the early 2000s. Indonesia is also promoting the domestic use of biodiesels by providing subsidies particularly because of its increasing demand for conventional fossil fuels and the subsequent need to import oil and gas. In the light of transportation and marketing costs as well as environmental footprint, it would be far smarter to use them domestically, rather than to export them to developed nations. Therefore, the "energy independence" discourse, instead of "green innovation" discourse, supports the domestic production and uses of biofuels within the developing nations (i.e., the same logic applies to the US policy for domestic production and use of bioethanol).

On the other hand, the frustrating experience with *Jatropha curcas* in many South Asian nations suggests the need of reframing its position in the varieties of biofuel options. *Jatropha* was once promoted as a method of increasing the biofuel production in arid areas where palm and other plantations are relatively difficult. The promotion of *Jatropha*, however, has been unsuccessful in many parts particularly because of the unstable demand for biofuel feedstock as well as the frustrating yields compared to what had been promised in pitched promotion. In response, Chap. 2.2.2 articulates a more realistic strategy for *Jatropha curcas*. Households in the rural parts of Southeast Asian nations are still suffering from the shortage of basic needs, including fuels. Instead of letting them cut down trees without much concerns on sustainability, *Jatropha curcas* could be useful in sustaining the life of rural villages by providing sustainable fuels for household.

Table 4.1 Stakeholders covered in each chapter of Part II

		Producers in developing nations	Users in developing nations	Producers and users in developed nations	Communities of stakeholders in the production areas	Future generations
2.1	*Impacts at the global scale*					
2.1.1	Impacts of biofuels to the international agricultural markets	+		++		
2.1.2	Effect of biofuel production on the GHG emission reduction	+				++
2.2	*Impacts at the national and regional scales*					
2.2.1	Changes in forest and land/water usage patterns following biofuel productions	++			++	+
2.2.2	Socioeconomic impacts in East Asia	+	+		+	+
2.3	*Impacts at the local scale*					
2.3.1	Political and societal impacts to the variety of stakeholders	++	+		+	
2.3.2	Impacts to the ecosystem services				++	

4.3 Producers and Users in Developed Nations

Chapter 2.1.1 is a unique, but foremost important, chapter in Part II, because it will primarily deal with producers and users of biofuel in the United States. While Brazil would be the first successful nation to propagate the use of biofuel through its Pró-Álcool policy, the renewed interest in biofuels in the twenty-first century was initially triggered by the US federal government's substantial investments in the further use of biofuels produced by domestic corns and soybeans. Its influence is formidable because using these feedstocks for biofuels directly competes with other conventional uses, which are vegetable oil and food as such. The added demand for these crops can trigger price hikes infiltrated by opportunistic investments in future option markets. In addition, wide varieties of government subsidies to producers, often motivated by political interests, in the name of "green innovation" distort the value of these crops. Nonetheless, the bioethanol production in the United States has been steadily increasing even until 2011[1], and the troubling nature of biofuels that entertain competition between fuel use and food use can become a major issue in 2012 when the North American farmers are hit by a major drought.

Users in the developing nations are also major stakeholders because they can influence the demand for biofuels worldwide. Chapter 2.3.1 will touch on this issue. In particular, the EU member states and many states of the United States have mandates regarding the mix of biofuels in the conventionally marketed automobile fuels. For instance, EU's Directive 2009/28/EC mandates each member state to turn the 10% of its transportation fuels into biofuels before 2020. This kind of mandate influences the global demand for biofuels. Users in developed nations are also concerned about rainforests and fair trade. Therefore, the governments of these nations have been exploring the use of accreditation schemes for biofuels so that their policy for increasing the use of biofuels would not harm the interests of these domestic NGOs and other interest groups.

4.4 Communities of Stakeholders in the Production Areas

There are many "other" key stakeholders in the field of production. For instance, Chap. 2.2.1 articulates the impact of increased production of feedstocks on the environment. In Brazil, there is strong concern, particularly among international environmental NGO communities, about the expansion of plantations into rainforest and in Cerrado. Even if a marginal expansion of plantation takes a piece of relatively less environmentally valuable land, it can have a spillover effect on water resources and other competing land uses such as cattle herding. These indirect impacts must be addressed in considering the expansion of biofuel uses and Chap. 2.2.1 tries to address these issues by analyzing such impacts quantitatively. Chapter

[1] http://www.ethanolrfa.org/pages/statistics

2.3.2 provides an overview of similar impacts from the perspective of ecosystem services.

Natural environment and resources are not only the key stakeholders related to production. For instance, Chap. 2.3.1 provides an overview of the relevant stakeholders in Brazil and Indonesia. Investors and trade firms play an integral role in developing the supply chain of biofuels. In Brazil, the national development bank, BNDES, plays a pivotal role in developing advanced facilities that can flexibly produce both crude sugar and bioethanol. Trade farms are also important in facilitating infrastructure developments for exporting biofuels at a large scale. Without an appropriate involvement of these stakeholders, the expansion of biofuel uses, particularly at the global scale, is unlikely. Labor organizations are also important. Plantations hire a number of manual seasonal laborers for harvesting. Once the biofuels are exported to developed nations, international communities will be more concerned about the working environment and "fair" share of profit between the plantation owners and laborers.

4.5 Future Generations

The last, but requiring a serious attention, category of biofuel deployment stakeholders is our future generations. The foremost goal of deploying biofuels at the global scale is to reduce the carbon emission, which will eventually curtail the risk of damage from a major climate change. Chapter 2.1.2 addresses this question by comparing various methods of biofuel production that can most reduce the GHG emission. While feedstock captures CO2 when it grows, the procedures of turning it into fuels in fact emit CO2. Chapter 2.1.2 therefore introduces the life cycle analysis perspective to measure the effect of various kinds of production method. In its analysis, the effect of using bagasse—the residue of sugar cane—for electricity production is substantial because the increasing demand for electricity in Brazil would lead to an increased dependence on coal-fired power plants.

4.6 Summary

This chapter reviewed how the stakeholder perspectives are applied to our analyses of biofuel deployment with different methods. While the discourse on biofuels has often focused on the impact of expanded production on the surrounding natural environment, the impact is far more extended to a wide variety of stakeholders. In reality, the issues around biofuel are not just a polarized debate between pro-expansion and anti-expansion. A number of actors, such as investors, manual laborers, and end-use consumers in developed nations, play a pivotal role in the chain of actions from production to consumption. In addition, political discourses often negate the foremost important stakeholders: the future generation. They are the ones who would eventually benefit from the curtailed carbon emissions. As is reviewed in this chapter, there is a clear need to draw a holistic picture of biofuel stakeholders in the field. In the next chapter, varieties of discourse over the biofuel uses are reviewed using the analytical framework called "ontology."

Part II
Impacts on Land Use and Ecosystem Services: Global Economic and Environmental Impacts

Chapter 5
Welfare Effects of the US Corn-Bioethanol Policy

Hideaki Takagi, Taro Takahashi, and Nobuhiro Suzuki

5.1 Introduction

Because of the surge in international crop prices in 2008, production of biofuel derived from crops has been criticized for expanding crop demands and threatening food security. In the USA, where corn is the main raw material for bioethanol, the demand for corn has rapidly increased from 18 million tons in 2001 to 100 million tons in 2008. Further, the Renewable Fuel Standard (RFS) included in the Energy Policy Act of 2005 requires refiners, blenders, and importers to use 36 billion gallons of renewable fuels by 2022, including more than 21 billion gallons of second-generation biofuels such as cellulosic ethanol. The use of corn as an energy source is expected to continue further expansion.

Many studies have simulated the crop price under biofuel production and measured its impact on the market equilibrium. For example, Koizumi and Ohga (2009) measured the impact of expansion of Brazilian FFV (flexible fuel vehicle) utilization and of the US biofuel policy on production, consumption, export, and import of sugar and corn.

However, their studies are confined to simulating the impact on market outcome. This leaves an important question: does higher price really reduce social benefit? It is sure that high crop price declines consumers' purchasing power and weighs upon their household economy. This is a critical issue, especially for low-income households. However, recent prices of agricultural commodities have been too low for farmers to sustain on. Many developed countries have scrambled to support them through production control and subsidies. Without these measures, farmers would be at a loss because of small revenue. In this regard, ethanol production can be

H. Takagi · T. Takahashi · N. Suzuki (✉)
Graduate School of Agricultural and Life Sciences, The University of Tokyo,
1-1-1 Yayoi, Bukyo-ku, Tokyo 113-8657, Japan
e-mail: asuzukiz@mail.ecc.u-tokyo.ac.jp

K. Takeuchi et al. (eds.), *Biofuels and Sustainability*, Science for Sustainable Societies, https://doi.org/10.1007/978-4-431-54895-9_5

regarded as one of the *solutions*. Expansion of demand for corn and its higher price will contribute to their revenues and reduction of governmental expenditures. It is misleading to judge for or against biofuel production with the fixed view that higher price is always harmful.

For these reasons, cost-benefit analysis should be carried out. In this study, we aim to find the most economically beneficial policies with regards to the US bioethanol production. The next section overviews the model structure of the US corn market incorporating bioethanol. In the third section, we will show the simulation results across five scenarios. In the fourth section, we will outline the method to calculate the benefits and costs to each stakeholder. Our conclusion is presented in the fifth section.

5.2 The Model Structure

5.2.1 *Overview of the Model*

The fundamental concept of our model used in this study is illustrated in Fig. 5.1. The left side of the chart represents the supply of corn, the middle part the demand for edible corn, and the right side the demand for ethanol. This model is a dynamic partial equilibrium model focused on US corn market.

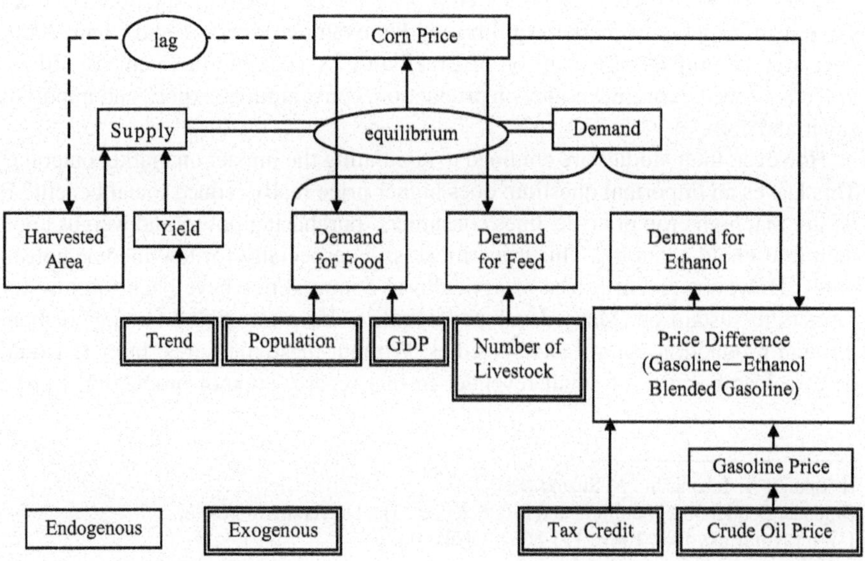

Fig. 5.1 Structure of the model

Farmers determine whether they cultivate corn or soybean before planting. If soybean price is relatively high and farmers expect soybean is more profitable, they plant soybean instead of corn. As a result, harvested area of corn will reduce. Similarly, demand for corn is also affected by wheat price because corn as feed can be substituted by wheat. We do not consider fluctuations of their prices to simplify the model's structure and interpretation of the result of our study. In this model, corn price is determined solely by US domestic supply and demand, and behavior of producers and consumers in other countries are not reflected in the price. Import is omitted from the model because it has been less than 0.22% of production since 1961 (FAOstat n.d.).

5.2.2 *Detailed Model Structure*

Equations are either estimated using data published by USDA (n.d.-a) and FAOstat (n.d.) or cited from Oga and Yanagishima (1996).

By the assumption mentioned above, corn supply consists of only the production in the year. Production "Q" can be divided into yield "Y" and harvested area "S":

$$Q = Y \times S, \tag{5.1}$$

where Y and S are represented, respectively, by

$$\ln Y = -2.61 + 1.09 \ln(T - 1921) \tag{5.2}$$

$$\ln S = -117.98 + 16.69 \ln T + 0.125 \ln P_{(-1)} + 0.083 \ln P_{(-2)} + 0.042 \ln P_{(-3)} \tag{5.3}$$

where "T" is a trend term equaling the calendar year. "P" is corn price, with (−1), (−2), and (−3) suggesting lagged variables.

According to the estimation result (5.2), the yield is not affected by corn price and increases as time passes. Equation (5.3) shows that harvested area is positively affected by past 3 years' corn prices and, ceteris paribus, expanding every year.

Corn demand can be divided into four different usages: for food, for feed, for bioethanol, and for export. "For food" means the corn directly consumed by people. The estimation result of demand for food per capita is

$$\text{Food} = \text{Food}_0 \left(\frac{\text{Pop}}{\text{Pop}_0}\right)\left(\frac{\text{P}}{\text{P}_0}\right)^{-0.21}\left(\frac{\text{GDP}}{\text{GDP}_0}\right)^{-0.2} \tag{5.4}$$

where "Food," "Pop," and "GDP" mean demand for food, population of the USA, and real GDP of the USA, respectively. Variables with subscript 0 are their actual values in 2005.

Demand for feed is described by the price of corn and livestock production. Livestock production includes beef, pork, mutton, chicken, egg, and milk. The estimation result of demand for feed in the USA is

$$\mathrm{Feed} = \mathrm{Feed}_0 \left(\frac{\mathrm{Beef}}{\mathrm{Beef}_0}\right)^{0.27} \left(\frac{\mathrm{Pork}}{\mathrm{Pork}_0}\right)^{0.11} \left(\frac{\mathrm{Chicken}}{\mathrm{Chicken}_0}\right)^{0.08} \left(\frac{\mathrm{Egg}}{\mathrm{Egg}_0}\right)^{0.10} \left(\frac{\mathrm{Milk}}{\mathrm{Milk}_0}\right)^{0.14} \left(\frac{\mathrm{P}}{\mathrm{P}_0}\right)^{-0.4} \tag{5.5}$$

where "Feed," "Beef," "Pork," "Chicken," "Egg," and "Milk" mean demand for feed, beef production, pork production, chicken production, egg production, and milk production, respectively. Variables with subscript 0 are actual values in 2005. All elasticities in Eqs. (5.4) and (5.5) are estimated by Oga and Yanagishima (1996). In their study, mutton production elasticity of demand for feed in the USA is shown to be insignificant.

The demand for bioethanol is expressed as follows. Since it is ethanol producers who purchase corn for ethanol, the demand function should represent the ethanol producer's behavior. But there is the final consumer's behavior to purchase ethanol behind their behavior. That is, if it is interpreted that bioethanol production is as much as consumption, the bioethanol producer's demand for corn reflects the final consumer's demand for bioethanol. Therefore, this model does not consider the bioethanol producer as an intermediary but the final consumer who wants "liquid corn" called bioethanol.

In the USA, bioethanol is sold by being added to gasoline. The standard and target rates of blending differ by states. In our model, we assume only two types of vehicle fuel: gasoline and E10. "Gasoline" in the equation indicates the pure gasoline made from crude oil. "E10" is blended gasoline which includes 10% of bioethanol in volume. Since there is no substantial difference between gasoline and blended gasoline as a vehicle fuel, consumers select which fuel to buy according to their own preference. Therefore, the demand for blended gasoline is supposed to depend on the price difference each consumer can accept:

$$\mathrm{Eth}/\mathrm{Pop} = -0.00530 + 5.50\times10^{-6}\times\mathrm{Pdif} + 2.67\times10^{-6}\times T \tag{5.6}$$

"Eth" means corn consumption for bioethanol production. Corn demand for bioethanol production per capita is explained in this equation. "Pdif" is the retail price difference:

$$\mathrm{Pdif} = P^{*}_{\mathrm{gas}} - P^{*}_{E10} \tag{5.7}$$

Both "P^{*}_{gas}" and "P^{*}_{E10}" represent their own retail prices per gallon. Consumers must convert these prices into those per mile in order to compare accurately their efficiencies because the heating value per gallon of ethanol is about 60% that of gasoline. Our estimations (5.6) showed, however, that the demand was explained better by the price difference per gallon than by that per mile. This was presumably because the heating value ratio of E10 to gasoline was calculated as 1 × 90 % + 0.6 × 10 % = 96%, and thus consumers did not care about such a small efficiency difference.

$$P^*_{\text{gas}} = P_{\text{gas}} + \text{fueltax} \tag{5.8}$$

where "P_{gas}" is the gasoline price before tax. This is apparently dependent on crude oil price "P_{p}" as shown in the following equation:

$$P_{\text{gas}} = -6581 + 3.317T + 1.745P_p \tag{5.9}$$

Similarly, retail price of E10 is

$$P^*_{E10} = P_{E10} + \text{fueltax} - \text{taxcredit} \tag{5.10}$$

"taxcredit" indicates the tax credit for a gallon of E10 that is deducted from federal fuel tax. This was 5.1 cent/E10gallon until 2008. The price of E10 before tax and deducted "P_{E10}" is represented as (5.11)

$$P_{E10} = P / 2.7 \times 10\% + P_{\text{gas}} \times 90\% \tag{5.11}$$

About 2.7 gallons of bioethanol is produced from a bushel of corn. The term *P*/2.7 in Eq. (5.11) means the raw material cost to produce a gallon of bioethanol. Since E10 consists of 10% of ethanol and 90% of gasoline, P_{E10} is calculated by weighted average. Although other costs such as transportation cost and margin of bioethanol producer are not considered here, we view that what is important in our model is not the level of the price difference but the change in the price difference. As the change in bioethanol price is almost explained by its raw material price, this allows us to omit these other costs.

Back to Eq. (5.6), bioethanol consumption per capita is explained well by the price difference and the trend term. Adding Eqs. (5.7), (5.8) and (5.9) we can clearly see that a rise in crude oil price brings a rise in gasoline retail price, then expansion of the price difference, and, finally, a higher E10 consumption. A rise in corn price diminishes bioethanol consumption in reverse. When the price difference is fixed, bioethanol consumption tends to increase as time passes.

The last part of the model is the demand for export. According to FAOstat (n.d.), the trend of the corn export of the USA has stopped at 45–50 million tons in recent 20 years although there are millions of tons of fluctuation. In addition, the corn production in the USA has reached 300 million tons. Therefore, we round its fluctuation to fix the export at 48 million tons:

$$\text{Ex} = 48000 \tag{5.12}$$

Overall, the demand function in total is expressed as

$$D = \text{Food} + \text{Feed} + \text{Eth} + \text{Ex} \tag{5.13}$$

Finally, at the equilibrium, it holds that

$$D = Q \tag{5.14}$$

5.3 Simulation

5.3.1 Overall design

We have to introduce some assumptions for our simulation analysis. Our simulation begins 2007 and ends at 2020. The corn market in 2007 and 2008 was in an unusual situation caused by unexpected factors such as the financial crisis. Since our model is recursive, including these noises prevents us to analyze the mainstream trend in the grain market. Thus we avoid 2009 as the initial year. In addition, this method allows us to see how unusual the actual situation was in 2008 by comparing the actual value with the equilibrium value which is solely determined by supply and demand.

Population, GDP, livestock productions, and crude oil price are exogenous to the model. For population and GDP, projected values from USDA (n.d.-b) are used. Livestock productions are simply explained by the trend term. Their trends through the simulation period are shown in Fig. 5.2. Crude oil price is assumed to rise by 2% a year.

Fig. 5.2 The US livestock production (projection) (Source: Estimation result using data from FAOstat (n.d.))

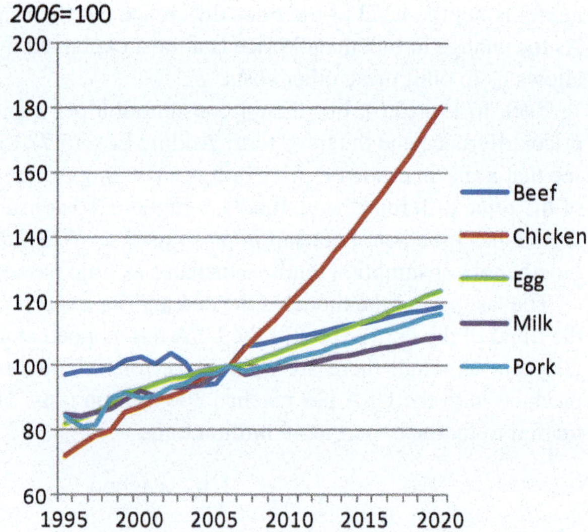

5.3.2 Scenarios

We arranged two types of scenarios. The first group consists of the baseline and four scenarios in which the tax credit is shifted as shown.

Baseline: *taxcredit* = 5.1 cent/E10 gallon (actual value in 2008)
Scenario 1: No Ethanol Production (NEP)
Scenario 2: *taxcredit* = 0 cent/E10 gallon
Scenario 3: *taxcredit* = 10 cent/E10 gallon
Scenario 4: *taxcredit* = 18.4 cent/E10 gallon (totally offsets the current federal fuel tax)

The policies in Scenarios 1 and 2 are expected to result in less bioethanol production than the baseline level, whereas those in Scenarios 3 and 4 are expected to result in reverse.

The second group of scenarios consists of the baseline and two scenarios in which E10 is replaced with E20. There is a crucial assumption here; the parameters in the bioethanol demand function (5.7) remains unchanged even if the blending rate has changed.

Baseline: *taxcredit* = 5.1 cent/E10 gallon (actual value in 2008)
Scenario 5: consumers select gasoline or E20 (low)
Scanario 6: consumers select gasoline or E20 (high)

Tax credit is 10.2 cent/E20gallon in both Scenarios 5 and 6. This is because *taxcredit* is a variable indicating the tax credit for E10. In other words, *taxcredit* = 5.1 means the tax credit for ethanol is 51 cent/gallon. If this rate is fixed, the one for E20 equals to 10.2 cent/gallon.

The difference between Scenario 5 and Scenario 6 lies in the interpretation of the bioethanol demand function (5.7). If "Eth" in this function is interpreted as the demand for bioethanol proper, that is, the amount of bioethanol in the E10 or E20 mix, the change from E10 to E20 does not alter the consumption because it is determined only by the price difference between gasoline and blended gasoline. This is Scenario 5. However, the demand function (5.7) can also be interpreted as the demand for *blended gasoline* because the consumption of bioethanol and that of blended gasoline are two sides of the same coin under our assumption. That is, the demand function for blended gasoline is identical regardless of the blending rate. The consumption of bioethanol in the E20 scenario is twice that of baseline if the price difference is the same. This is Scenario 6. These concepts are illustrated in Figs. 5.3 and 5.4. As the result, the demand function is altered as shown.

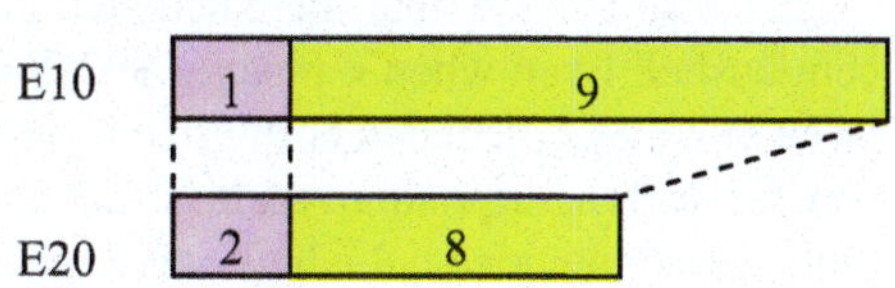

Fig. 5.3 Energy consumption in Scenario 5

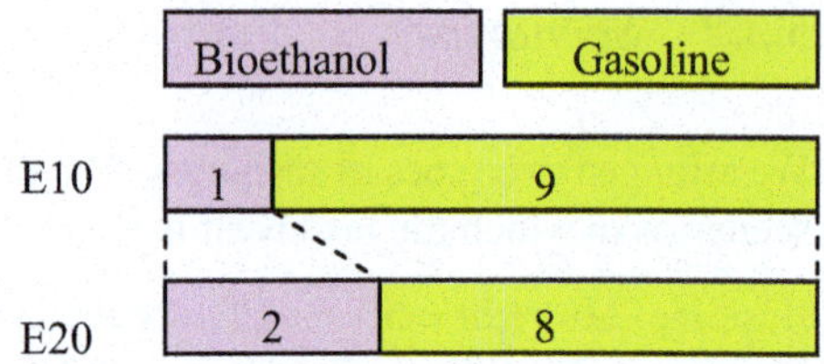

Fig. 5.4 Energy consumption in Scenario 6

Scenario 5:

$$P_{E10} = P / 2.7 \times 20\% + P_{\text{gas}} \times 80\% \left(12'\right)$$

Scenario 6:

$$\text{Eth} / \text{Pop} = \left(-0.00530 + 5.50 \times 10^{-6} \times \text{Pdif} + 2.67 \times 10^{-6} \times T\right) \times 2 \left(7''\right)$$

$$P_{E10} = P / 2.7 \times 20\% + P_{\text{gas}} \times 80\% \left(12''\right)$$

5.3.3 *Results*

Figure 5.5 shows the simulation results of corn price in the first group along with the actual values from 1991 to 2006. In all scenarios, the price goes downward. This is especially remarkable in NEP. In all scenarios but NEP, the price settles in the range of 200–300 cent/bushel, the level at which price has actually stayed for more than 30 years.

Expanding demand for corn is met mainly through growing yield. The US corn yield was 9.5 t/ha in 2007 and is expected to be 10.8 t/ha in 2020. Harvested area does not expand so much in any scenario. In 18.4 cent/gallon scenario which needs the largest area of all scenarios tested, it is expected to be 32.3 million ha in 2020. Although it exceeds the maximum area recorded prior to the simulation's initial year (30.4 million ha in 1985), the difference is not so large relative to its amplitude.

In 2008, the actual corn price jumped up to 497.5 cent/bushel (USDA/ERS n.d.-a). One of the causes was bioethanol production. Because crude oil price in 2008 was $97.26/bbl (BP n.d.), bioethanol consumption must have been promoted. Then, how much impact did the rise in crude oil price have on corn price? We simulated corn price in 2008 by setting crude oil price at the actual value.

The results show that the calculated corn price in 2008 is no more than 257.7 cent/bushel. Even when replacing $97.26/bbl with $147/bbl (the highest crude oil price in 2008), calculated corn price is only 306.8 cent/bushel. The most likely reason for such a surprising result is that a rise in crude oil price brings not only a higher gasoline price but a higher E10 price.

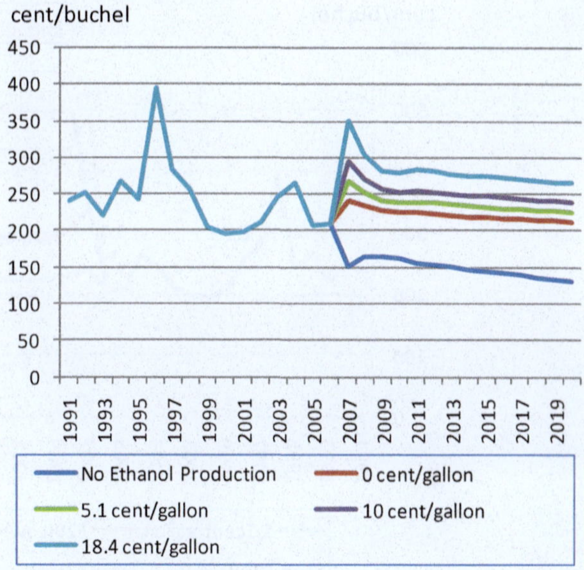

Fig. 5.5 Simulation result: the first group

This leaves about 240 cent/bushel of the difference that cannot be explained only by supply and demand. This component is caused by external factors such as the financial crisis and excessive expectation of investors.

The results of the second group simulations are shown in Fig. 5.6. The corn prices in these two scenarios are much higher than that in the baseline. The price in E20 (high) scenario (Scenario 6) almost reaches the actual value in 2008 and that in E20 (low) scenario (Scenario 5) becomes as high as the 18.4 cent/gallon scenario in the first group, but unlike the first group, they do not decline. The price in E20 (low) scenario stays at about 294 cent/bushel and that in E20 (high) scenario rises and reaches 477 cent/bushel in 2020.

According to these results, raising blending rate is expected to have a greater impact on corn price than increasing tax credit.

5.4 Welfare Analysis

5.4.1 Overall design

In the previous section, the simulation result of corn price was shown. On the basis of this result, we analyze who benefit by the US bioethanol policy in each scenario and by how much.

Six countries and one region are considered here: the USA, China, Argentina, Mexico, Brazil, EU (consisting of 27 countries), and Japan. All but Japan are main

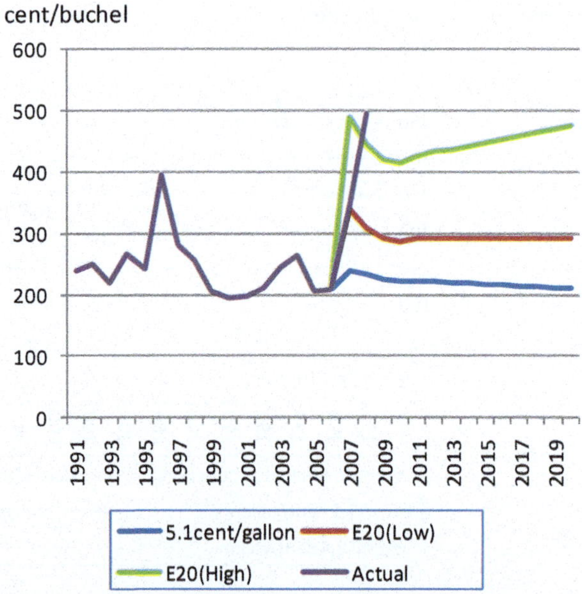

Fig. 5.6 Simulation result: the second group

corn producers in the world. Especially, the USA and China produce more than 100 million tons individually and account for about 60% of the whole production in the world between them. Of course, each producer is also a consumer. Japan produces little corn and hence is regarded as a sole consumer.

Corn producers and corn consumers in these countries are assumed to be economic stakeholders. Corn consumers refer to people who consume corn directly as food or indirectly as feed. They are all assumed to be price takers and behave based on an exogenously determined price.

The US government and the US consumers of *bioethanol* are also included as stakeholders. The US government deducts the federal *fueltax* for ethanol. Thus, even if bioethanol production improves social welfare, too much support increases the opportunity cost and incurs financial difficulties on the state.

There are other benefits of bioethanol such as CO_2 reduction, saving fossil fuels, improvement in energy self-sufficiency ratio, and prevention of air pollution. Although they are regarded as significant sources of positive externalities, there is no consensus on assessment method. Therefore, they are omitted in the subsequent analysis. The value of CO_2 reduction, however, will be discussed later using an approach different from the direct evaluation method.

5.4.2 Detailed Procedures

Benefits for corn producers and corn consumers are evaluated as producers' surplus and consumers' surplus, respectively. Supply function and demand function are necessary to calculate surplus in each country.

The structure of the corn demand in non-US countries is almost the same as that of the USA. The only difference is that it does not include a demand for bioethanol (Fig. 5.3). Here we show them in general form below. Subscript *i* indicates countries.

Demand for food:

$$\text{Food}_i = \text{Food}_{i0}\left(\frac{\text{Pop}_i}{\text{Pop}_{i0}}\right)\left(\frac{\text{GDP}_i}{\text{GDP}_{i0}}\right)^a\left(\frac{P}{P_0}\right)^b$$

Demand for feed:

$$\text{Feed}_i = \text{Feed}_{i0}\left(\frac{\text{Cattle}_i}{\text{Cattle}_{i0}}\right)^c\left(\frac{\text{Pig}_i}{\text{Pig}_{i0}}\right)^d\left(\frac{\text{Sheep}_i}{\text{Sheep}_{i0}}\right)^e\left(\frac{\text{Poultry}_i}{\text{Poultry}_{i0}}\right)^f$$
$$\left(\frac{\text{Egg}_i}{\text{Egg}_{i0}}\right)^g\left(\frac{\text{Milk}_i}{\text{Milk}_{i0}}\right)^h\left(\frac{P}{P_0}\right)^j$$

Demand in total:

$$D_i = \text{Food}_i + \text{Feed}_i$$

The values of elasticity (superscripts *a-j*) are sourced from Oga and Yanagishima (1996). Livestock productions are assumed to be exogenous to the system.

Supply:

$$S_i = Q_i = Q_{i0}\exp\left(k(T-2006)\right)\left(\frac{P_{(-1)}}{P_{0(-1)}}\right)^m$$

where *k* and *m* are also the parameters peculiar to each country.

Producers' surplus and consumers' surplus in country *i* are calculated by integrating the supply function and the demand function, respectively. In order to allow the result converge and compare them among the scenarios, each surplus is expressed as the differential between the baseline scenario and the concerned scenario.

The relative producers' surplus:

$$\Delta PS_i = \int_{p\text{II}}^{p\text{I}} S_i \mathrm{dP}$$

The relative consumers' surplus:

$$\Delta CS_i = -\int_{p\mathrm{II}}^{p\mathrm{I}} D_i \mathrm{dP}$$

where P^{I} and P^{II} indicate the corn price in the baseline scenario and in the concerned scenario, respectively. Note that both supply function and demand function are fixed across the scenarios.

The benefit to ethanol consumers is also calculated as ethanol consumers' surplus. The surplus is calculated with

$$\mathrm{CS}_{\mathrm{eth}} = \int_{\mathrm{pi}}^{\mathrm{pe}} \mathrm{Eth\ dP}$$

The lower limit of integral "p^e" is the equilibrium price, and the upper limit "p^i" is the intercept of the inverse demand function. Therefore, the relative surplus to that of the baseline is derived with the following formula.

The relative bioethanol consumers' surplus:

$$\Delta CS_{\mathrm{eth}} = \int_{\mathrm{pi}}^{\mathrm{pe}} \mathrm{Eth\ dP} - \int_{\mathrm{piI}}^{\mathrm{peI}} \mathrm{Eth}^{\mathrm{I}}\mathrm{dP}$$

where the second term in this equation is the bioethanol consumers' surplus at the baseline.

The last to consider is the opportunity cost of the US government. As mentioned above, the US government loses the tax revenue by deducting the federal fuel tax. Since the tax credit increases as the government promotes the bioethanol production, the negative effect on the government is larger in such scenarios. Although rise in the corn price provides a positive aspect to the government of reduction of the agricultural subsidies, this effect is not included in this study.

The assumption at calculating the opportunity cost is that the domestic energy consumption in the given year is constant across the scenarios.

Suppose that consumption of gasoline is V gallon and that of ethanol is W gallon in a certain year. They are equivalent to V + 0.6 W gallon of gasoline in terms of energy since the heating value ratio is gasoline/ethanol = 1:0.6. Therefore, the tax revenue would be (V + 0.6 W) × *fueltax* if there were no ethanol consumption. The actual tax revenue is V × *fueltax* + W × (*fueltax*-10*taxcredit*). The opportunity cost is the differential between them. It is calculated with

$$\Delta \mathrm{Gov}_0 = W \times \left(10\mathrm{taxcredit} - 0.4\mathrm{fueltax}\right)$$

This value shows the loss of revenue comparing with NEP. To compare with baseline, we use the following equation:

$$\Delta \mathrm{Gov}_0^I = W^I \times \left(10\mathrm{taxcredit}^I - 0.4\mathrm{fueltax}\right),$$

where superscript *I* indicates the baseline. The relative loss in the concerned scenario compared with baseline is calculated with the relative loss for the government:

$$\Delta \mathrm{Gov} = \Delta \mathrm{Gov}_0 - \Delta \mathrm{Gov}_0^I$$

We are aware that including ΔGov in social welfare is debatable, and whether it should for part of the analysis or not depends on to whom this revenue is ultimately attributed. If the revenue of the US government gained by reducing tax credit is used for corn producers, corn consumers, or bioethanol producers, ΔGov should be included. Otherwise, all amount of ΔGov should not be necessarily included. We assume that all revenues are used exclusively for those associated with corn markets, hence include all of ΔGov in our calculations.

The total relative benefit of China, Argentina, Mexico, Brazil, and EU is evaluated as $\Delta \mathrm{PS}_i + \Delta \mathrm{CS}_i$. Those of the USA and Japan are $\Delta \mathrm{PS}_u + \Delta \mathrm{CS}_u + \Delta \mathrm{CS}_{\mathrm{eth}} + \Delta \mathrm{Gov}$ and $\Delta \mathrm{CD}_{\mathrm{j}}$, respectively.

5.4.3 Results

The results for 2020 are shown in Table 5.1. The unit is million US$.

Among the first group, the total welfare of the world is the largest in the 0 cent/gallon scenario. Figure 5.7 is the scatter plot between *taxcredit* and the sum of benefits. This figure shows that the sum is likely maximized at *taxcredit* = 0 under the constraint that *taxcredit* ≧0. The result that 0 cent/gallon scenario brings more total benefit to the whole world than NEP implicates the significance of bioethanol production. When the US benefit is excluded from the total, NEP brings the maximum benefit among five scenarios.

For all scenarios, the values of the US subtotal are 0 or less. This result stands consistently from 2008 through 2020. This means the actual US policy in 2008 has economic rationality. Figures 5.8 and 5.9 are the scatter plots between tax credit and the US benefit in 2020. These figures show the US benefit is maximized when *taxcredit* = 3.6, the level very close to the current level of 5.1 cent/gallon.

The US producers' surplus is more subject to the bioethanol policy than consumers' surplus. Therefore, $\Delta \mathrm{PS}_u + \Delta \mathrm{CS}_u$ is positive when the government adopts pro-bioethanol policy. Such policy also improves $\Delta \mathrm{CS}_{\mathrm{eth}}$. ΔGov, however, is decreased considerably, whereas NEP brings relatively small gain. As the result, maximization of the total US benefit is accomplished at the intermediate tax credit.

The benefit for Argentina becomes larger as bioethanol production is promoted, while that of China, Brazil, and Mexico decreases. In Japan, the benefit is necessarily decreased because of the assumption that there is no corn producer. The benefit of EU does not have simple trend. NEP is expected to bring more benefit than the current situation. EU has insisted that the bioethanol production derived from crops

Table 5.1 Evaluation result: benefits (unit, million US$; year, 2020)

			First group				Second group	
Country	Benefit	NEP	0 ¢	5.1 ¢	10 ¢	18.4 ¢	E20 (low)	E20 (high)
USA	ΔPS	−11,988	−1793	0	1852	5346	9381	35,981
	ΔCS	8745	1159	0	−1159	−3252	−5583	−18,478
	ΔC Seth	−16,554	−3503	0	3730	10,928	6193	12,490
	Δgov	3499	4043	0	−4747	−14,826	−1627	−4444
	Subtotal	−16,297	−94	0	−324	−1804	8364	25,549
CHN	ΔPS	−8474	−1268	0	1310	3783	6663	25,641
	ΔCS	11,487	1613	0	−1639	−4665	−8125	−28,897
	Subtotal	3013	346	0	−329	−881	−1463	−3256
ARG	ΔPS	−548	−91	0	96	286	513	2271
	Δ CS	225	30	0	−30	−85	−147	−492
	Subtotal	−323	−61	0	66	201	367	1779
BRZ	Δ PS	−2046	−328	0	346	1015	1809	7636
	Δ CS	4206	564	0	−566	−1594	−2745	−9222
	Subtotal	2160	237	0	−221	−579	−936	−1586
MEX	Δ PS	−888	−145	0	153	452	809	3494
	Δ CS	3008	411	0	−415	−1172	−2028	−6974
	Subtotal	2119	266	0	−261	−720	−1220	−3479
EU	Δ PS	−2727	−446	0	473	1396	2500	10,869
	Δ CS	2840	366	0	−363	−1011	−1723	−5508
	Subtotal	113	−80	0	110	385	777	5361
JPN	Δ CS	241	31	0	−30	−84	−142	−445
	Subtotal	241	31	0	−30	−84	−142	−445
World	Δ PS	−26,671	−4070	0	4230	12,278	21,674	85,893
	Δ CS	30,752	4174	0	−4203	−11,864	−20,493	−70,016
	Δ C Seth	−16,554	−3503	0	3730	10,928	6193	12,490
	Δ Gov	3499	4043	0	−4747	−14,826	−1627	−4444
Total		−8973	645	0	−990	−3483	5747	23,923
Excluding the USA		7324	739	0	−666	−1679	−2617	−1626

should be abandoned. Their argument is that the higher food price brought by bio-ethanol production would cause hunger to the poor. The evaluation result shows that it is rational for EU itself as well. However, it also gains more benefit in the 5.1 cent/gallon scenario than in NEP, and the benefit increases as tax credit becomes larger.

E20 scenarios in the second group are expected to bring larger benefits to the world. However, almost all of them belong to the USA, especially the US corn producers. This means the E20 policy might cause a greater gap in international distribution of the benefit; on the one hand, the USA gains enormous benefit, but, on the other hand, China loses their share. Domestic gaps in distribution also tend to be wider under these scenarios; for example, in China under E20 (high) scenario, corn consumers have to tolerate a great loss, while corn producers gain a large benefit. Such scenarios will likely be unacceptable unless benefit transfer is implemented.

Fig. 5.7 Tax credit and total sum of benefits

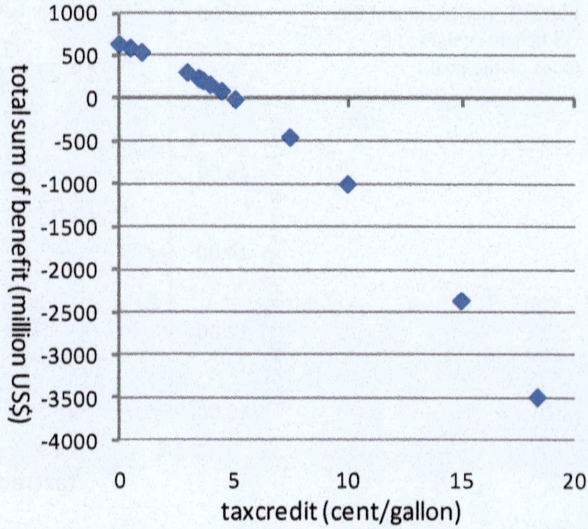

Fig. 5.8 Tax credit and the US benefit

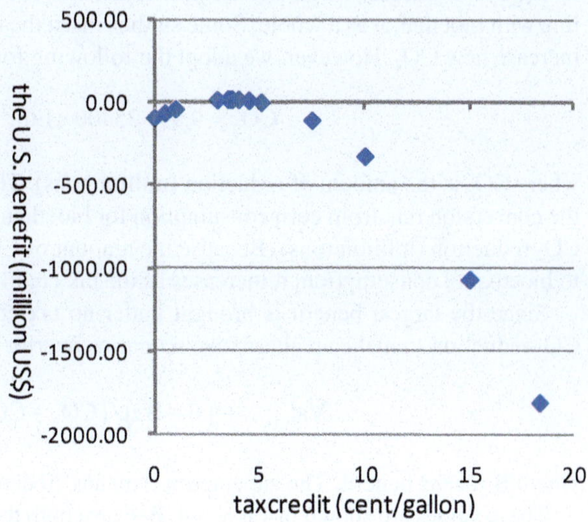

5.4.4 *Value of CO_2 Reduction*

In the previous subsection, it was shown that the most rational E10 scenario differs for the whole world (0 cent/gallon) and the USA (3.6 cent/gallon). Those values, however, include only economic factors and for others are omitted. Now we attempt to introduce the value of CO_2 reduction. As it is difficult to evaluate the value of CO_2 reduction, we will apply another approach.

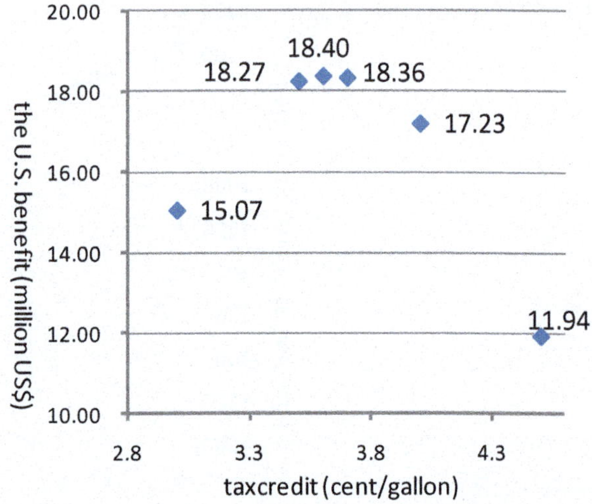

Fig. 5.9 Tax credit and the US benefit (enlarged to focus on the peak)

Table 5.2 shows the volume of CO_2 reduction. Bioethanol also emits CO_2 through its life cycle. There is no consensus how much CO_2 is reduced by substituting gasoline with bioethanol as a whole. Some studies insist the use of bioethanol rather than increase new CO_2. However, we adopt the following formula here.

$$CO_2 = 9.99 / 25400 \times \text{Eth},$$

where CO_2 is the amount of reduction (million tons). The coefficient 9.99/25,400 is the conversion rate from corn consumption for bioethanol production (1000 tons) to CO_2 reduction (million tons). Because the amount of CO_2 reduction is proportional to bioethanol consumption, it increases as the tax credit becomes larger (Table 5.3).

Since the largest benefit is brought under no tax credit, the threshold value of CO_2 reduction to make an alternative scenario superior is calculated as

$$\text{Val}_{CO_2} = -\left(B - B^0\right) / \left(CO_2 - CO_2^0\right),$$

where B means benefit. The superscript 0 means "0 cent/gallon" scenario.

These values are shown in Table 5.4. Because both the benefit and the CO_2 reduction of NEP are less than those of the 0 cent/gallon scenario, NEP can never exceed the 0 cent/gallon scenario. Therefore, the result is described as "inferior (-)." The value for "Average" in Table 5.4 is calculated as

$$\text{Average} = \sum_{2020}^{t=2011} B_t \Big/ \sum_{2020}^{t=2011} CO_{2t}$$

The smallest average is \$116.9/CO_2t in the 3.6 cent/gallon scenario. This means 3.6 cent/gallon scenario is more rational than any other scenarios for the world if the value of the CO_2 reduction is evaluated to be greater than \$116.9/t. In other words,

Table 5.2 CO_2 reduction (unit: million tons)

	NEP	0 ¢	3.6 ¢	5.1 ¢	10 ¢	18.4 ¢
2011	0	17.6	19.8	20.7	23.6	28.5
2012	0	18.5	20.7	21.6	24.6	29.6
2013	0	19.5	21.7	22.6	25.6	30.7
2014	0	20.4	22.7	23.6	26.6	31.7
2015	0	21.4	23.7	24.6	27.6	32.8
2016	0	22.4	24.7	25.6	28.7	33.9
2017	0	23.4	25.7	26.6	29.7	35.0
2018	0	24.4	26.7	27.7	30.8	36.1
2019	0	25.4	27.8	28.8	31.9	37.3
2020	0	26.5	28.9	29.8	33.0	38.4

Table 5.3 Total benefit of the whole world. Unit: million US$

	NEP	0 ¢	3.6 ¢	5.1 ¢	10 ¢	18.4 ¢
2011	−5106	289	122	0	−607	−2347
2012	−5649	274	117	0	−585	−2267
2013	−6103	298	125	0	−615	−2368
2014	−6483	345	139	0	−668	−2531
2015	−6843	397	155	0	−724	−2698
2016	−7251	442	169	0	−772	−2837
2017	−7674	488	183	0	−821	−2985
2018	−8100	537	198	0	−875	−3145
2019	−8532	590	215	0	−932	−3312
2020	−8973	645	231	0	−990	−3483

Table 5.4 Minimum value of CO_2 reduction required to make the scenario the global optimum. Unit: US$/ton

	NEP	3.6 ¢	5.1 ¢	10 ¢	18.4 ¢
2011	–	77.2	94.4	149.8	241.3
2012	–	71.7	88.4	141.9	230.0
2013	–	78.4	95.2	149.2	238.3
2014	–	91.9	108.9	163.8	254.5
2015	–	107.2	124.4	179.8	271.5
2016	–	119.7	137.0	192.7	284.9
2017	–	132.5	149.9	205.9	298.8
2018	–	146.1	163.7	220.2	313.9
2019	–	160.4	178.1	235.0	329.5
2020	–	175.0	192.8	250.2	345.4
Average	–	116.9	134.2	189.8	281.9

the most rational policy for the USA becomes acceptable by the world. Note, however, that we use the word "acceptable" in a narrow sense that the most rational tax credit for the world is not necessarily 3.6 cent/gallon even in this case.

5.5 Conclusion

In this study, we simulated corn price under the current and alternative sets of bioethanol policy and then analyzed the social benefit associated with each scenario.

First, the simulation result showed the following:

(a) Corn price in any scenario will decline even if crude oil price rises 2% a year.
(b) Although too much support for bioethanol production might induce higher corn price than the usual level, the current policy of tax concession will contribute to support the price. On the contrary, suspension of bioethanol production might cause price slashing.
(c) Switching E10 to E20 has a much larger impact on corn price than changing the level of the tax credit policy.
(d) The hike in corn price in 2008 is scarcely explained by supply and demand only, which indicates that the major cause was not expansion of bioethanol production but external factors. Thus, although bioethanol production induced excessive expectation of investors, it will unlikely persist.

The second step of our study aimed to measure the impact of the US bioethanol production on household economy. The result of this step showed the following:

(a) The current policy of tax concession (5.1 cents) is at a rational level for the US society.
(b) The USA is expected to gain another 11.3 million dollars of benefit (average of 2011–2020) by reducing the tax credit to 3.6 cent/gallon, the theoretical maximum.
(c) Ethanol production without tax credit brings most beneficial result for the whole world combined.
(d) The value of CO_2 reduction must be more than \$116.9/t in order for the 3.6 cent/gallon scenario to become acceptable to the world.
(e) Although the E20 policy might produce much more benefit for the world than the tax credit policy, the distribution of benefit will likely to be less equal than the current situation.

Overall, three observations can be made in relation to the present analysis.

First, the most rational policy is not exactly the same as the most appropriate policy. Any policy change generates winners and losers both internationally and domestically. The problem can be solved if benefit transfer is carried out successfully, however it is very difficult especially to dissolve international inequality. It is a critical problem when low-income countries or such households become "losers" even if the policy is the most efficient for the whole world or for the USA.

Second, an increase in producers' surplus does not necessarily mean improvement of famers' revenue. In case when there is imbalance of market power among farmers, wholesalers, and retailers, it is possible that the surplus does not come back to farmers at all. Even if it is not the case, their surplus is partially offset when expansion of bioethanol utilization is caused by a surge in crude oil price. This is because higher crude oil price means an increase in production cost of corn as well as dominance of blended gasoline against pure gasoline. Therefore, part of their additional revenue flows out as an additional cost. A higher price in agricultural commodities does not immediately benefit farmers in many cases.

Finally, we have to consider the impact on individuals. More specifically, we need additional studies at domestic level to answer the following questions.

(a) Are there any ways to transfer their benefit appropriately?
(b) How much loss can each stakeholder tolerate?
(c) Who, ultimately, receives the benefit?
(d) How much impact does policy impose on each household?

References

BP Global (n.d.) http://www.bp.com/bodycopyarticle.do?categoryId=1&contentId=7052055. Last access 27 Jul 2010

FAOstat (n.d.) http://faostat.fao.org/. Last access 27 Jul 2010

Koizumi T, Ohga K (2009) Impact of the expansion of Brazilian FFV utilization and U.S. biofuel policy amendment on the world sugar and Corn Markets: an econometric simulation approach. Jpn J Rural Econ 11:9–32

Oga, Yanagishima (1996) International food and agricultural policy simulation model

USDA/ERS (U.S. Department of Agriculture/ Economic Research Service) (n.d.-a) Feed grains database http://www.ers.usda.gov/Data/FeedGrains/. Last access 27 July 2010

USDA/ERS (U.S. Department of Agriculture/ Economic Research Service) (n.d.-b) International macroeconomic data set http://www.ers.usda.gov/Data/Macroeconomics/. Last access 27 July 2010

Chapter 6
The Effect of Biofuel Production on Greenhouse Gas Emission Reductions

Keisuke Hanaki and Joana Portugal-Pereira

6.1 Introduction

Fossil fuel consumption is a major cause of climate change. Biofuels can reduce the consumption of fossil fuels and thus reduce carbon dioxide emissions, because biofuels are carbon neutral. More specifically, the carbon dioxide that is emitted when a biofuel is burned merely returns to the atmospheric carbon dioxide that was taken into plants from the atmosphere by photosynthesis. Therefore, biofuels seem to be a very effective means for reducing these emissions, at least at first sight.

However, the reality is not so simple but controversial (Edwards et al. 2007; Fargione et al. 2008; Hill et al. 2006; Menichetti and Otto 2009; Searchinger et al. 2008). The production of a biofuel consists of growing an energy crop and using biomass obtained from the crop as a raw material for making liquid fuel. The production of bioethanol includes processes of fermenting sugarcane, corn, or other sugar-based feedstock and distilling the contents in a similar manner to distilling liquor, and the production of biodiesel includes a chemical reaction (transesterification) using vegetable oil as a raw material. Additionally, feedstock is collected in the farmland to the fuel production plants, and then biofuels are distributed to filling stations, where they are sold to consumers. These processes inevitably consume energy.

Moreover, it has been pointed out that greenhouse gases (GHGs) are also generated by the cultivation of energy crops. One of the GHGs, nitrous oxide, is generated when fertilizers are used to raise the yields of energy crops. Furthermore, when forest land is converted to use for energy crop plantations, the carbon dioxide absorption of the forests is lost, in addition to which organic matter in the soil breaks down and generates carbon dioxide. All these issues mean that the production of

K. Hanaki (✉) · J. Portugal-Pereira
Department of Urban Engineering, The University of Tokyo,
7-3-1 Hongo, Bunkyo, Tokyo 113-8656, Japan
e-mail: hanaki@iniad.org

K. Takeuchi et al. (eds.), *Biofuels and Sustainability*, Science for Sustainable Societies, https://doi.org/10.1007/978-4-431-54895-9_6

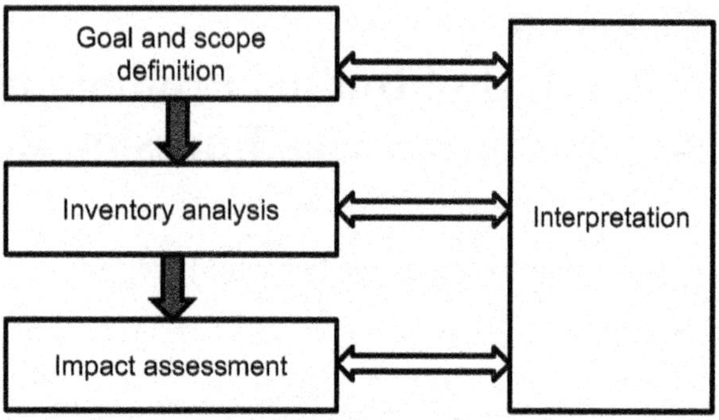

Fig. 6.1 Basic scheme of life cycle assessment

biofuels leads to GHG emissions, to a greater or lesser extent. In other words, there is a trade-off between the carbon dioxide reductions when biofuels are used instead of fossil fuels and the GHGs that are generated in the production of biofuels.

A quantitative evaluation with life cycle assessment (LCA) is necessary for a judgment of this trade-off. The environmental loading of a biofuel can be evaluated from the beginning to the end use in LCA. The net effect may be evaluated by calculating increases and reductions in GHGs throughout the life cycle.

LCA is a tool originally devised for the chemistry industry. Their salient features are combining and quantitatively evaluating environmental loads associated with the manufacture of a product, from the acquisition of raw materials to disposal. LCA is useful in design for environment and employed for sustainable consumption and consumption of products with low environmental loading. The environmental loading associated with creating and disposing of these products may not be immediately apparent. This hidden environmental loading is estimated by LCA and is referred to as embodied environmental loading. Terms such as embodied energy and embodied CO_2 are used when evaluating the environmental aspects of products. The prime example of application of LCA for the public is carbon footprint, a figure focusing on carbon dioxide.

The basic phases of LCA are shown in Fig. 6.1. The environmental loads include energy consumption and emissions of a wide range of substances such as GHGs, air pollutants, water pollutants, or heavy metals. First, the goals and scope of an analysis are specified. In this phase, the scope is suitably determined with consideration to the purpose for using LCA, and the environmental loads to be analyzed are specified. Next, an all-encompassing inventory of emissions relating to the selected environmental loads is created. In the impact assessment phase, these environmental loads undergo a comprehensive evaluation. The environmental loading is ascertained, and processes are improved by interpretation phases.

Previous researches on LCA of biofuels have been controversial and produced diverse reports. In some reports, biofuels were found to be beneficial for the environment, and in other reports, they were found to be ineffective or even damaging. Reasons for the contradictions among these results include the questions of which environmental loads were considered and which biofuels were considered. Another issue is that environmental loading differs between different geographic regions.

Although there are different carbon intensity of the fuels that are used in ordinary manufacturing industries, GHG emissions are broadly proportional to energy consumption. However, this does not apply to biofuels. Byproducts and residues of biomass cultivation processes and production processes are carbon-neutral biomass. When these byproducts and residues are used as energy sources, this counts as energy consumption but does not count as GHG emissions. Therefore, life cycle energy consumption and life cycle GHG emissions must be evaluated differently. LCA could indicate that biofuel does not save energy but does lead to a reduction in GHGs.

6.2 Biofuel LCA

6.2.1 LCA Framework

When biofuels are used as a fuel for vehicles, LCA is applied in a way that enables comparison with a LCA for conventional fuels, such as gasoline or diesel fuels. When these fossil fuels are used, crude oil is extracted from an oil well, transported, refined, sold, and loaded into a vehicle fuel tank. These steps are referred to as the well-to-tank (WTT) stage. The fossil fuel is then burned while the vehicle is running, which is referred to as the tank-to-wheel (TTW) stage. An analysis of the two together is referred to as a well-to-wheel (WTW) analysis. It is convenient to use this division for biofuels also for comparison with fossil fuels.

GHG emissions from fossil fuels are small in the WTT stage and large in the TTW stage. In contrast, with a biofuel that is a carbon-neutral fuel, the environmental loading is small for the TTW stage and large for the WTT stage, complicating the analysis.

Although biofuels are not obtained by digging oil wells like petroleum, the WTT stage consists of raw material acquisition, processing, storage, and distribution (Fig. 6.2).

The raw material acquisition stage includes growing an energy crop. Analysis of the energy loading of this cultivation step is an analysis of the energy loading for agricultural activity, which differs from the industrial production for which LCA is usually used. For industrial production, the raw materials and energy that are input, and the pollutants generated by the artificial manufacturing are ascertained. These are originally understood at the design stage for the artificial manufacturing process, so data and reports are easy to obtain. In contrast, agriculture depends on nature, and effects on the environment that result from artificial utilization of land must be evaluated. Unlike an industrial process that is performed under controlled conditions, the environmental loading of agriculture has to be evaluated under conditions that are greatly influenced by climate and the like. The fate of fertilizers that are used and changes in the soil that are caused by agricultural activities should also be included.

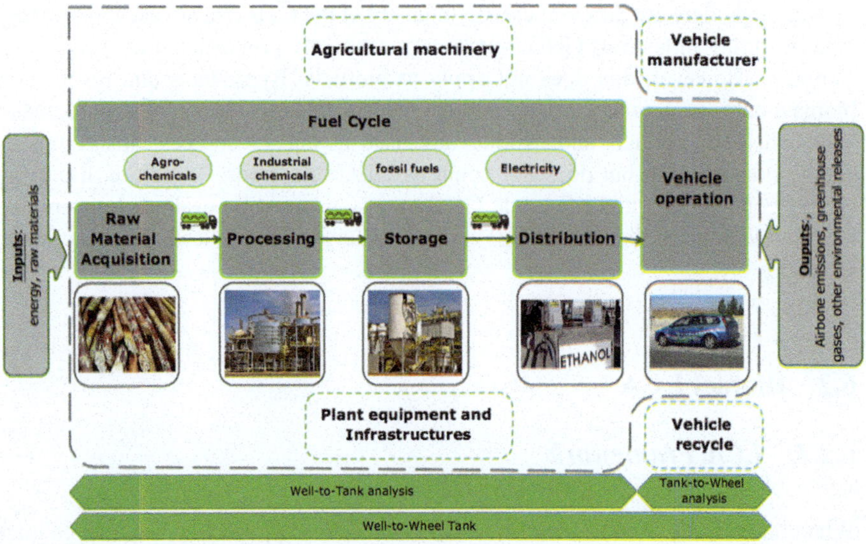

Fig. 6.2 Well-to-tank and tank-to-wheel analysis of biofuel

In GHG emissions, there is the possibility of large amounts of nitrous oxide being generated during cultivation of feedstocks. This originates from nitrogen fertilizers and is produced by processes of nitrification and denitrification. The amounts generated do not depend solely on the amounts of nitrogen fertilizer used but also on soil characteristics and climate conditions, so there can be a very wide range of values for emission factors. It should be noted that when land with a soil such as peat that stores large amounts of organic carbon is developed and cultivated, the carbon accumulated in the soil is released into the atmosphere as carbon dioxide. The carbon dioxide is generated from a natural source, but it should be accounted as anthropogenic production.

The production of biofuels is an industrial process, so it does not differ greatly from the usual LCA. However, there are several points to consider. Because factories that carry out these production processes are often located in agricultural areas, available energy sources in those locations are limited. Unlike industrial raw materials, energy crops include many unusable parts such as straw and husks, and how these byproducts are used is an important question.

In the TTW stage, combustion conditions differ from the case of using standard gasoline or diesel. Therefore, differences in vehicle combustion efficiency and differences in atmospheric pollutant emission are important and will vary depending on the performance of the vehicles using the fuels.

When each phase of LCA is applied to biofuels, the key issues are shown in Fig. 6.3. In the goal and scope definition phase, target energy crop, agricultural waste production, and process must be determined. The comparison of different kinds of environmental loading is the main issues in the impact assessment phase.

In LCA functional unit (FU) is determined first, and environmental loading is calculated with reference to these functional units. For example, in an LCA for

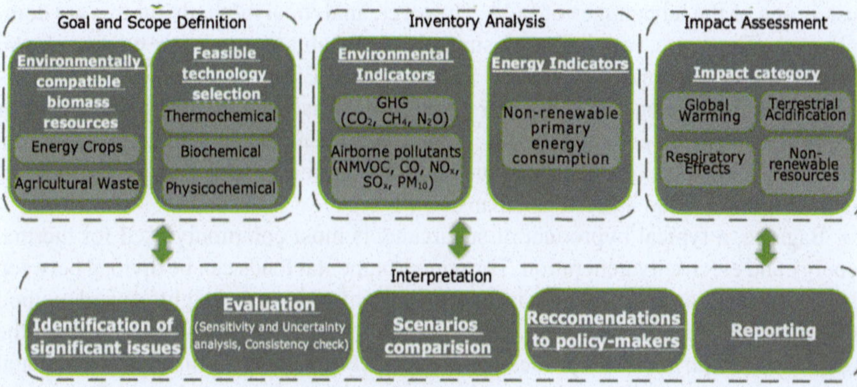

Fig. 6.3 Key issues in each step of LCA of biofuel

refrigerators, one refrigerator of certain volume is the functional unit, and for a raw material such as cement, the functional unit is 1 ton of cement. Comparing results that are organized by the same functional unit is a fundamental feature of LCA. Different functional units may be used depending on the reasons for carrying out LCA.

For biofuels, there are a number of possible different functional units that can be considered. Cherubini and Strømman (2011) identified four different FUs that are commonly used to describe bioenergy systems: (i) feedstock based, which refers to the unit of input biomass and does not depend on conversion processes and products; (ii) product based, which allows an evaluation from a downstream angles; (iii) agricultural land related that refers to the farmland used to cultivate feedstock; and (iv) per year unit that assesses results in a yearly basis. A simple example of a functional unit is 1 l of the biofuel. However, if gasoline and bioethanol are compared, the different fuels have different heat values, so a simple comparison by units of volume is inappropriate. One method often used to express results is units of energy of biofuel (e.g., kJ), which is reasonable for evaluating biofuel production processes. However, if the final TTW stage is included, the performance actually obtained is running a vehicle. Therefore, the results of the LCA may be expressed in vehicle kilometers traveled (VKT). This method considers biofuels from the downstream end of the material flow. It is also possible to look at the upstream end and express the environmental loading by production volumes of biofuel per hectare of agricultural land.

6.2.2 Evaluation of Effects of Byproduct Recovery and Use

One of the characteristics of biofuels is that large amounts of byproducts and residues, such as straw and husks, are produced in the processes of cultivation and production. It is important for the LCA to evaluate whether these byproducts and residues are used effectively. The possibility of quantitatively evaluating effective use is one of the advantages of LCA. However, analytical methods for this are complicated. There are two different strategies to evaluate apportion environmental loads between the final products and its byproducts. Either practitioners can proportionally divide the burdens between output flows based on their physical (mass or energy) content or economic value (called as allocation) or alternatively system boundaries can be expanded to include additional credits related to the byproducts displacement (called as system expansion).

Bagasse, a typical byproduct of sugarcane, is most commonly used for thermal energy and electricity generation. Thus, environmental loads can be divided between bagasse and ethanol based upon its energy content. This is a straightforward method that guarantees stable outcomes, as physical properties are constant. However, the allocation of physical properties may encounter criticisms, since environmental loads are not necessarily proportional to products' mass/energy content. Thus, practitioners can consider the expansion of the system boundary. When bagasse is used as an energy source, it has the effect of replacing a fossil fuel. In the LCA, this effect is evaluated as the avoided environmental loading from the fossil fuel. This is a typical case of system expansion.

If petroleum is replaced by a byproduct, the effect may be calculated as a reduction in the environmental loading caused by petroleum production. When electricity is generated by a byproduct, the effect reduces the environmental load associated with generating electricity. The substance replaced by a byproduct depends on the geographic region, and because a number of different kinds of fuel are used in a power grid, it is important to determine which fuels are replaced by the byproduct. For example, if electricity is generated by a byproduct in a region where the grid electricity is generated with coal, this leads to a large reduction in CO_2, whereas if electricity is produced from a byproduct in a region that uses hydroelectric power generation or nuclear power generation in the grid, the CO_2 reduction effect is very low.

As an example, we consider composting solid waste and substituting it for a chemical fertilizer (Fig. 6.4). In this case, the system boundary must be expanded to include fertilizer production. If the waste were not recycled through composting, the waste would be landfilled, causing an environmental load such as methane emission, while energy would be used in the industrial production of chemical fertilizer. Various environmental loads are caused by these processes. On the other hand, if compost is produced from the waste, an environmental load is caused by this production, but the abovementioned environmental loads of landfilling the waste and producing the chemical fertilizer are avoided. These calculations must be performed so that the effectiveness of the compost matches the effectiveness of the chemical fertilizer that is replaced. As the fertilizing effects of 1 ton of compost and 1 ton of chemical fertilizer are not the same, adjustment is necessary in the calculation.

With biofuels, energy substitution by residues and replacement of chemical fertilizers are typical uses of byproducts. Large amount of residues are produced from the conventional process for producing a biofuel from an energy crop. This residue can replace large amount of fossil fuel. On the other hand, if the production yield is raised by an innovative process, less amount of residue can be used to replace fossil

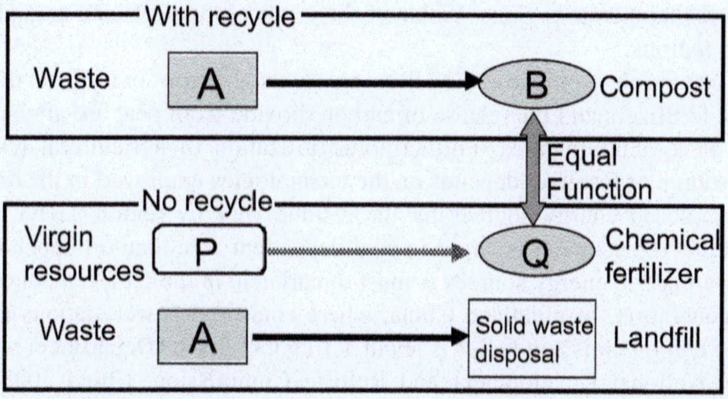

Fig. 6.4 LCA of waste utilization

fuel. As a result, the environmental loading per unit of energy for the biofuel produced may be smaller in the conventional process than in the innovative process. In other words, the innovative process can produce more biofuel, but environmental load per unit fuel increases.

Not readily biodegradable materials such as wooden stems are decomposed by pretreatment and then converted in the so-called second-generation biofuel production. As a result, residues used for process energy in the first-generation biofuel process are reduced in the second-generation process. Apparently environmental loading per amount of biofuels is larger, though yield of biofuels from the energy crop is high in the second-generation process. This is the controversial issue of the functional unit.

6.2.3 *The Necessity of Localized LCA*

Discussions of how local characteristics are included in LCAs and whether they should be included involve profound questions. Universal evaluations irrespective of location are often pursued in manufacturing LCA. Moreover, because the components and materials used in ordinary industrial processes are distributed all over the world, it is not practical to identify the production locations and include local characteristics of the production locations in LCAs. Therefore, it is appropriate to use global average values.

With biofuels, however, long-distance transportation of energy crops is not practical, and all stages up to production are carried out in the region of cultivation. Therefore, the environmental loading associated with biofuel production is affected by the characteristics of each region. First of all, the selection and growth yield of an energy crop greatly depends on the climate of a region. Sugarcane, cassava, oil palm, and similar crops grow in tropical regions, while other energy crops are suitable for temperate regions. Yields of the crops per unit of land area also differ between regions.

GHG emissions from the soil, such as emissions of nitrous oxide from the applied nitrogen fertilizers and the release of carbon dioxide from peat are affected by the soil of an agricultural area. Furthermore, utilization of agricultural residues as energy source or fertilizer depends on the technologies employed in the region.

The replaced energy sources by the residue vary by region. GHG reduction through this replacement is larger in coal-dependent area than oil-dependent area. This difference in energy sources is most remarkable in the replacement of electric power generation. In mainland China, where coal-fired power stations are dominant, the carbon emission factor is about 1.1 kg CO_2/kWh (Department of Climate Change, National Development and Reform Commission, China 2008). In the Tokyo region of Japan, the carbon emission factor in 2010 was about 0.4 kg CO_2/kWh (Ministry of Environment 2010), only one-third as much. This means that replacement of power generation would have a large CO_2 reduction effect in China and a small effect in the Tokyo region. The amounts would be even smaller in a region that uses hydroelectric power, such as Brazil (0.2 kg CO_2/kWh) (Portugal 2011).

The TTW stage is not affected by the production region but by local characteristics of regions in which a biofuel is consumed. Large amounts of biofuel are distributed internationally. Average lifespan and performance of vehicles and atmospheric pollution standards differ greatly between countries. Mixing ratios for biofuels and conventional fuels also vary between countries. For instance, in Brazil, ordinary gasoline (commonly referred to as *gasohol*) is blended with 18–25% (v/v) of anhydrous ethanol (MAPA 2011b), whereas in Japan the legal limit of ethanol blends is 3% (v/v) (Fukuda et al. 2006).

These points show that, when evaluating the GHG reduction effect of biofuels, LCA must be carried out considering the local characteristics of the producing regions and consuming regions.

6.3 Sugarcane Ethanol Production in Brazil

As stated earlier, LCA is a useful tool to evaluate the climate change mitigation potential of biofuels. Yet, it is also a source of controversy as LCA results are significantly dependent on local conditions of production and utilization, and options made by practitioners when selecting system boundaries, allocation procedures, and the functional unit of the system, among others. Thus, the truthful GHG and fossil fuel resource savings from biofuel life cycle and uncertainty factors behind LCA results are yet to be surely understood. To clarify these matters, a LCA has been conducted to evaluate the GHG emission and nonrenewable energy (NRE) consumption of sugarcane ethanol production in the South-Center region of Brazil and its application in the Brazilian national passenger vehicles. The analysis is focused on current practices, taking as reference the base year 2008 (the latest year for which

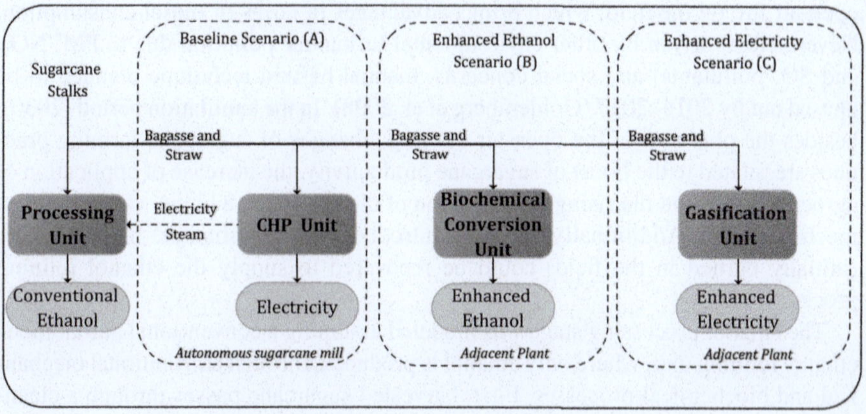

Fig. 6.5 Schematic diagram of baseline and alternative scenarios

data inventory was available), as well as on the forecast of potential technological and efficiency improvements up to a 2030 horizon. Results are presented through different angles, in terms of function unit selection and allocation procedures, in order to understand how background assumptions and methodological choices constrain the overall LCA results.

6.3.1 Description of the Case Study and its Scenarios

Brazil is an important World player in the ethanol market, being the second largest producer after the USA. Nearly 9.67 million hectares were dedicated to sugarcane farming, producing around 25.7 million m^3 of ethanol (MAPA 2011a). While the production of sugarcane ethanol is a well-known and optimized process, significant gains can be achieved by enhancing the efficiency recovery of its byproducts, bagasse, and straw. In this view, a baseline scenario and two alternative scenario forecasts were considered. The baseline scenario (A) attempts to describe the current state of the art of sugarcane farming practices and ethanol refining units, as well as likely future improvements, without considering any processing technological shift. Alternative routes, on the other hand, forecast the recovery of bagasse and straw via advanced biochemical or thermochemical processes, designed as cellulosic ethanol scenario (B) and enhanced electricity scenario (C), respectively. The scheme of the baseline and alternative scenarios is presented in Fig. 6.5.

Accordingly, the baseline scenario encompasses the *status quo* of the sugarcane farming and ethanol processing activities. Sugarcane farming occurs in a 6-year cycle, including five harvest periods with gradual yield decline and one planting season. During the farming stage, fossil fuel energy consumption is mainly associated with agrochemical application and diesel used in machinery during agricultural operations. Harvest activities are primarily manual and labor intensive, followed by an open-air fire on the field, which brings advantages in terms of diesel consumption. Nevertheless, driven by other environmental (urban air pollution due to PM, NO_x, and SO_x pollutants) and social concerns, manual harvest technique planned to be phased out by 2014–2017 (Goldemberg et al. 2008). In the simulation period (2030), besides the phaseout of the open-air burning, changes of sugarcane farming practices are related to the boost of sugarcane productivity, the increase of application of agrochemicals, and the rising consumption of diesel (due to a higher rate of harvest mechanization). Additionally, with the introduction of mechanical harvest, straw (initially burned in the field) could be recovered to supply the ethanol refining process.

The ethanol processing stage was modeled assuming a conventional autonomous ethanol refining unit, where only ethanol is produced, through conventional mechanical and biochemical processes. First, harvested sugarcane passes through a cleaning unit to remove impurities, followed by an extraction system, where sugarcane is chopped, and shredded, and juice with high content of sugar is separated and cleaned. Bagasse and filter cake are also generated as coproducts. Following juice extraction, the mixture is fermented by yeasts (commonly the *Saccharomyces cerevisiae*). Finally, the resulting wine is purified through fractional and azeotropic distillation processes. Besides the final product anhydrous ethanol, vinasse is also generated. As this coproduct has a high nutrient content (N, P, K), it is commonly recovered and used for ferti-irrigation. One tone of vinasse recovers 0.36 kg of N-fertilizer (Donzelli 2007).

The ethanol processing consumes energy for activating pumps, fans, and milling equipment, as well as thermal energy for the juice concentration and distillation processes. The process is assumed to be energy self-sufficient, i.e., the consumed energy is entirely powered by bagasse and straw (from mechanically harvested fields), in combined heat and power (CHP) units. Currently, CHP systems are generating steam at low pressure (~22 bar), which results in limited electricity generation. However, old boilers are being replaced by efficient high-pressure steam boilers (~65 to 90 bar, 480 °C) that increase the amount of surplus electricity (Macedo et al. 2008; Seabra et al. 2010). Thus, in the baseline scenario, forecasts up to a 2030 horizon were modeled taking into consideration the shift to high-pressure steam boilers and penetration of more efficient processes in ethanol production.

Alternatively, the cellulosic ethanol route (scenario B) considers the integration of an adjacent plant next to the principal ethanol distillery unit that produced cellulosic ethanol (the so-called second-generation ethanol) sourced by disposed bagasse and sugarcane straw. Prior to the fermentation and purification stages, the pretreatment processes are applied. Acid or enzymatic hydrolysis is done in order to separate degradable cellulose and hemicellulose compounds from the nondegradable lignin compounds. Accordingly, bagasse and straw biomass are pretreated via diluted sulfuric acid, followed by enzymatic hydrolysis with co-fermentation. The product is recovered, and purification follows common processes of the sugarcane-derived ethanol. Thus, an extra 46.3 l per ton of sugarcane is expected to be generated from the cellulose coproducts.

Table 6.1 Product flows of baseline and alternative sugarcane ethanol production scenarios (2030 horizon)

	Outputs	
	Ethanol	Electricity
Scenarios	$m^3.ha^{-1}$	$MWh.ha^{-1}$
Baseline (A)	9.6	14.0
Cellulosic ethanol (B)	14.3	5.2
Enhanced electricity (C)	9.6	18.7

On the other hand, the enhanced electricity route (scenario C) assumes that the disposed bagasse and sugarcane straw are recovered in a gasification unit. Biomass is firstly dried, conditioned, and then transformed into syngas. Later, syngas generates electricity in a gas turbine combined cycle (GTCC). The thermal and electrical efficiency of GTCC units is higher than conventional CHP units; thereby electricity production of enhanced electricity scenario is optimized. This option gives clear priority to electricity production, whereas enhanced ethanol route gives advantage to ethanol production.

Table 6.1 summarizes the main characteristics of these scenarios. Despite the common input flows, different processes are applied to each of the alternative scenarios. Thereby, output flows are considerably divergent. Predictable, in cellulosic ethanol route, ethanol production is maximized (14.3 m^3 per ha, namely, 54% higher than in the baseline scenario), whereas in electricity route, surplus electricity generation is prioritized (18.6 MWh per ha, which is 29–33% more than in the baseline scenario).

With regard to the utilization stage, three different light passenger vehicles were modeled, according to the Brazilian fleet. In Brazil, three different vehicles can be found: (i) E25 Otto-cycle vehicles running with E25 blended fuel (25% of anhydrous ethanol and 75% of conventional gasoline), referred to as gasohol; (ii) flexible-fuel vehicles (FFV), a new technology of vehicles that detects in real time the ratio of oxygen and fuel in the engine and accepts any kind of ethanol/gasoline blend (In this study, a share of 60% of hydrous ethanol and 40% of gasohol is assumed.); and (iii) dedicated ethanol vehicles that are exclusive only on hydrated ethanol (E100) that were discontinued in 2007.

6.3.2 LCA Framework

In this study an LCA framework was applied, which encompasses the goal and scope definition, the inventory analysis, life cycle interpretation assessment, and interpretation of result consistency. Figure 6.6 illustrates the steps followed in the LCA conducted.

The following paragraphs describe in detail the most relevant parameters and assumptions taken into consideration when evaluating the sugarcane ethanol production and utilizing scenarios.

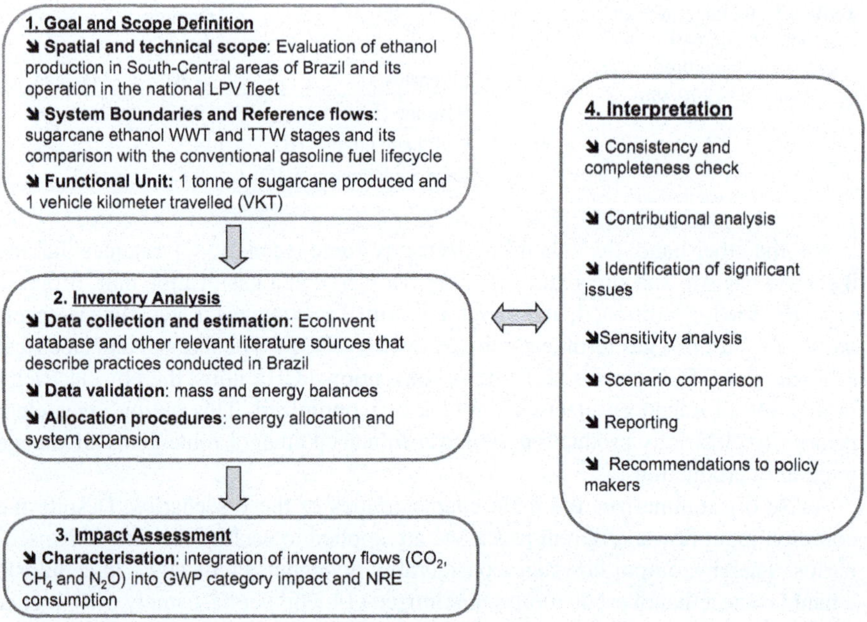

Fig. 6.6 Theoretical framework applied in this study

6.3.2.1 Definition of System Boundaries and Reference System

The evaluation included both the WTT stage that encompasses the sugarcane farming, ethanol refining processes, and intermediary transportation stages (collection of sugarcane and distribution of ethanol) and the TTW stage, which reflects the utilization of ethanol blended fuel in the E25, FFV, and E100 vehicles. The designed sugarcane ethanol scenarios have been compared with the equivalence reference system. Once ethanol is potentially substituting conventional gasoline, the reference system chosen is the production of gasoline and its use in conventional Otto-cycle light passenger vehicles. Figure 6.7 presents the system boundaries of sugarcane ethanol scenarios and the reference systems.

6.3.2.2 Functional Unit

Two different FUs were selected for this study: (i) a product-based unit, i.e., 1 vehicle kilometer traveled (VKT), which considers the efficiency of the ethanol blended fuel combustion in the E25, FFV, and E100 vehicles, and (ii) a feedstock-related unit, 1 ton of sugarcane harvested, which reflects solely the production of sugarcane and is independent from the ethanol processing and operation stages.

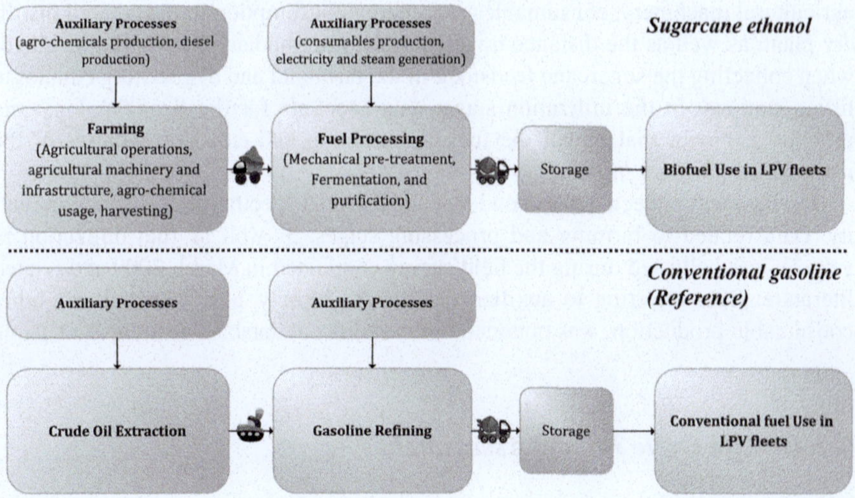

Fig. 6.7 System boundaries of sugarcane ethanol life cycle and reference system (conventional gasoline life cycle)

6.3.2.3 Allocation Procedures

Two different approaches have been followed to divide the environmental loads and energy consumption between ethanol and its byproducts. Thus, energy-based allocation and system expansion approaches were selected. In the energy-based allocation, environmental burdens were allocated based on the energy content of each of the products and byproducts. Accordingly, in the farming phase, loads have been portioned between sugarcane stalks and straw, over 98% of the loads being associated to the former. In the fuel production stage, loads were divided between ethanol, bagasse, and filter cake. Being ethanol the final product of the system, allocation factor accounts for nearly 78% in the baseline scenario (2030). As for the enhanced ethanol scenario, the allocation factor is larger, as it assumes maximization of ethanol production. On the contrary, in the enhanced electricity pathway, the allocation factor is lower because higher amount of electricity is produced than in the baseline scenario.

Alternatively, the system expansion approach assumes that the surplus electricity produced from the bagasse and sugarcane straw recovery displace Brazilian electricity grid. Thereby, 1 kWh of surplus electricity substitutes kWh of grid electricity.

6.3.3 Life Cycle Inventory Analysis

The life cycle inventory analysis provides the necessary input and output flows to model the environmental burdens associated to the sugarcane ethanol production and utilization in vehicle. As for the production stage, data include agrochemical production and usage in the farmland, diesel consumed during farming activities, agricultural machinery, consumables and energy consumption in the ethanol distillery plant, as well as the distance traveled in the intermediary transportation stages when collecting the sugarcane feedstock in the farmland and distributing ethanol to filling stations. In the utilization stage, data accounts for the fleet typology and mobility behavior that reflect the fuel consumption and emission patterns of the national vehicle fleet in Brazil.

The data was collected based on local and regional specificities of Brazilian reality. Data related to farming and processing stages, as well as fuel utilization in vehicle, was collected during the field survey conducted in March 2009 and related literature. Data referring to auxiliary processes, namely, agrochemical and other consumable production, was obtained in the ecoinvent database (Stutter 2006).

6.3.4 *Life Cycle Impact Assessment*

In the life cycle impact assessment (LCIA), the GHG emission and nonrenewable energy (NRE) consumption impact categories have been evaluated. The GHG is calculated as carbon dioxide equivalents (CO_{2e}), being carbon dioxide along with methane and nitrous oxide the greatest anthropogenic contributors. Global warming potential for 100-year time horizon in IPCC (2006) was used in Eq. 6.1:

$$\text{GHG} = \sum f_i . m_i \tag{6.1}$$

where:

GHG Emissions of CO_{2e} [mass unit]

f_i Characterization factor (global warming potential), 1 for CO_2, 25 for CH_4, and 298 for N_2O (IPCC 2006)

m_i Emissions of CO_2, CH_4, and N_2O [mass unit]

NRE consumption accounts for the amount of fossil fuels withdrawn during the life cycle of a product. It is given as the ratio of primary fossil fuel energy required throughout the alternative fuel life cycle, as shown in Eq. 6.2:

$$\text{NRE} = \frac{E_{\text{xp}}}{E_f} \tag{6.2}$$

where:

NRE Nonrenewable energy consumption ratio ($MJ.MJ^{-1}$)

E_{xp} Input primary fossil fuel energy required during the life cycle of the product

E_f Final energy output

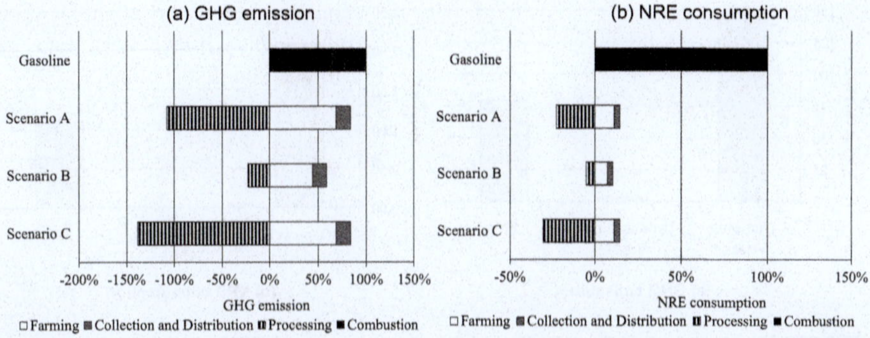

Fig. 6.8 Comparison of sugarcane ethanol production scenarios with gasoline fossil fuel production (2030)

According to the definition (Malça and Freire 2006), if the NRE consumption ratio is less than 1, it means that the fuel is renewable. On the other hand, if the ratio is larger than 1, more fossil energy is required to make the fuel than the energy available in the final fuel product. Thus, the fuel is classified as nonrenewable.

The life cycle inventories of the WTT part reflect the GHG emission and NRE consumption during the production stage. The results suggest that the baseline ethanol production pathway saves GHG emissions and NRE expenditure even without considering the carbon neutrality of biofuels, as shown in Fig. 6.8 in which gasoline combustion is shown as 100% for the comparison. The alternative route that prioritizes the generation of electricity from the bagasse and straw (scenario C) leads to greatest savings in terms of GHG and NRE consumption, owning to the displacement of the grid electricity by the surplus electricity generated. Once the surplus electricity generated from bagasse and straw releases nearly null GHG emissions and NRE expenditure, it presents lower impacts than Brazilian grid electricity. By this mean, net credits are obtained because the impacts during processing stage are regarded as negative. The net GHG and NRE consumption impacts of enhanced electricity production route are 155% and 116% less than gasoline, respectively. As against this, the production of enhanced ethanol (scenario B) pathway yields lower direct emissions than scenario C, owning to its higher yield of ethanol per hectare, but generates less electricity than the electricity route because the bagasse and straw waste are less.

The major steps that contribute to environmental loads are farming activities including fertilizers and agrochemical use and diesel consumption by tractors and other agricultural machineries. The fuel processing step is a modest contributor or even results in GHG and NRE consumption credits, since the energy in the ethanol refinery is supplied by bagasse and straw, which are renewable and carbon-neutral resources. The feedstock collection and fuel distribution steps also play a minor role in the overall environmental performance of the ethanol routes.

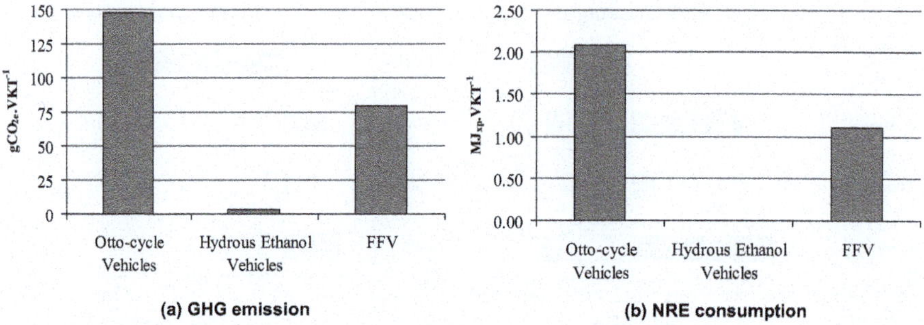

Fig. 6.9 Comparison of E25 and E100 vehicle operation in Brazil (2030)

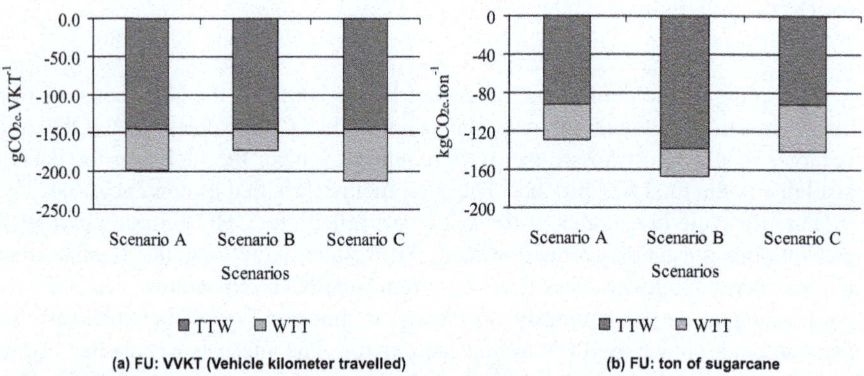

Fig. 6.10 Influence of FU in the GHG of WTW sugarcane ethanol scenarios

The TTW analysis of the operation of ethanol in conventional Otto-cycle (E25), ethanol-dedicated (E100), and flexible-fuel vehicles (FFV) shows a clear advantage of ethanol fuel in terms of GHG emission and NRE consumption, as displayed in Fig. 6.9. In fact, GHG emissions and NRE consumption from the combustion of ethanol are admitted to be null, given that ethanol is a carbon-neutral and renewable fuel. Thus, in E100 vehicles, GHG emissions only account for CH_4 and N_2O pollutants, which are nearly negligible. The environmental impacts of FFV vehicles solely reflect the 40% share of gasoline.

The WTW analysis integrates the previously displayed results of WTT and TTW analysis. Figure 6.10 presents the avoided GHG emissions and NRE consumption of ethanol production pathways and its use in E100 vehicles. The FUs are VKT and weight of sugarcane in Fig. 6.10a, b, respectively. All results show negative values which mean that GHG emission is reduced in all cases.

The analysis based on different FU apparently shows different results. On the one hand, the enhanced electricity scenario (scenario C) that maximizes the production of electricity via the bagasse gasification results in larger savings of NRE expenditure and GHG emissions per VKT (Fig. 6.10a), due to displacement of grid electricity. On the other hand, the enhanced ethanol scenario (scenario B) that shows the lower generation of surplus electricity results in the lowest savings. However, through a feedstock-oriented FU (Fig. 6.10b), results reveal different perspectives. The enhanced ethanol scenario (scenario B) that prioritizes the production of ethanol via biochemical synthesis of bagasse seems to be the most advantageous when applying a feedstock-based FU, as it shows higher yield of ethanol production per ton of sugarcane. Despite the lower savings by the use of small amount of waste biomass, the enhanced ethanol scenario shows larger savings during the operation stage, as it has higher yields of ethanol production per ton of sugarcane, than scenarios A and C. Therefore, savings per ton of sugarcane are more significant than in the other evaluated routes.

6.4 Final Remarks

The potential of biofuels to mitigate climate change and reduce dependency on fossil fuels is involved in an intense controversy, as its real benefits are significantly constrained by local geographic factors, technology of production, background assumptions, and methodological parameters of LCA. Major sources of uncertainty are data inventory, selection of allocation procedures, system boundaries, and functional unit. An LCA was conducted to discuss the source of uncertainty in LCA and to evaluate the GHG emission and NRE consumption category impacts of sugarcane ethanol production and utilization in Brazil within a 2030 horizon.

The results suggest that ethanol carriers effectively yield GHG and NRE savings, both in the production and operation stages. In the production stage, a key advantage is the recovery of sugarcane byproducts, straw, and bagasse, either to maximize the production of ethanol or to prioritize the generation of electricity. The former has lower direct emissions, but the latter results in GHG and energy credits as generated surplus electricity displaces grid electricity in Brazil. In the operation stage, the use of ethanol either in conventional, ethanol-dedicated, or flexible-fuel vehicles results in negligible GHG emission and NRE consumption, as ethanol is admitted to be a carbon-neutral renewable fuel.

The integrated WTW analysis discloses the overall benefits of ethanol carriers. Applying both a product-based and feedstock-related FU, ethanol shows gains in terms of GHG emission and NRE consumption, but results have dual interpretation according to which FU is selected. When applying a VKT as FU and system expansion approach, the enhanced electricity route reveals higher credits. On the contrary, a ton of sugarcane FU indicates that the enhanced ethanol pathway brings more advantages. This implies that the better process choice depends on the purpose and evaluation criteria.

This study has shown a wide variation of GHG emission and NRE consumption results, depending upon the selection of the functional unit, allocation procedures, and biofuel technology production pathways. Thus, it calls the attention to the need of improving LCA framework in order to evaluate the sources of uncertainty in complex systems, such as biofuel life cycles.

References

Cherubini F, Strømman AH (2011) Life cycle assessment of bioenergy systems: state of the art and future challenges. Bioresource Technol 102:437–451

Department of Climate change, National Development and Reform Commission, China (2008) "2008 Baseline Emission Factors for Regional Power Grids in China" obtained on Dec, 2009 at http://cdm.ccchina.gov.cn/english/

Donzelli JL (2007) Use of fertilisers. In: Macedo IC (ed) Sugarcane's Energy: twelve studies on Brazilian sugarcane agribusiness and its sustainability, 2nd edn. Berlendis & Vertecchia: UNICA, Brazil, São Paulo

Edwards R, Larivé J-F, Mahieu V, Rouveirolles P (2007) Well-to-Wheels analysis of future automotive fuels and powertrains in the European context. Well-To-Tank Report Version 2c, Tank-to-Wheels Report Version 2c, and Well-To-Wheel Report Version 2c, (CONCAWE, EuroCar and IES publication, EUR 22342 EN, Luxembourg, EU. Available from: http://ies.jrc.ec.europa.eu/

Fargione J, Hill J, Tilman D, Polasky S, Hawthorne P (2008) Land clearing and the biofuel carbon debt. Science 319:1235

Fukuda H, Kingsbury A, Obara K (2006) Japan bio-fuels bio-fuels production report. GAIN Report Number: JA6024. USAD

Goldemberg J, Coelho S, Guardabassi P (2008) The sustainability of ethanol production from sugarcane. Energy Policy 36(6):2086–2097

Hill J, Nelson E, Tilman D, Polasky S, Tiffany D (2006) Environmental, economic, and energetic costs and benefits of biodiesel and ethanol biofuels. Proc Natl Acad Sci U S A 103:11206

IPCC (2006) IPCC: Guidelines for National Greenhouse gas inventories, Prepared by the National greenhouse gas inventories programme

Macedo IC, Seabra JEA, Silva JEAR (2008) Green house gases emissions in the production and use of ethanol from sugarcane in Brazil: the 2005/2006 averages and a prediction for 2020. Biomass Bioenergy 32:582–595

Malça J, Freire F (2006) Renewability and life cycle energy efficiency of bioethanol and bio-ethyl tertiary butyl ether (bioETBE): assessing the implications of allocation. Energy 31(15):3362–3380

MAPA (2011a) Statistic Annuarium of Agroenergy 2010 (*Anuário estatístico de agroenergia 2010*) (*in Portuguese*) (http://www.agricultura.gov.br/desenvolvimento-sustentavel/agroenergia/publicacoes) (June 2011)

MAPA (2011b) Provisory Measure n°532 (*Medida provisória n*°532, de 28 de Abril de 2011) (*in Portuguese*) (http://www.jusbrasil.com.br/legislacao/1027558/medida-provisoria-532-11) (June 2011)

Menichetti E, Otto M (2009) Chapter 5: energy balance & greenhouse gas emissions of biofuels from a life-cycle perspective. In: Howarth RW, Bringezu S (eds) Biofuels: environmental consequences and interactions with changing land use, Germany, (SCOPE 2009)

Ministry of Environment, Japan (2010) CO2 emission factor of electricity (in Japanese): http://www.env.go.jp/earth/ghg-santeikohyo/material/denkihaishutu.html

Portugal J (2011) Life cycle assessment of the alternative fuel production and utilisation in light passenger vehicle fleets in Brazil and India. Doctoral dissertation. Faculty of Engineering, The University of Tokyo

Seabra JEA, Tao L, Chum HL, Macedo IC (2010) A techno-economic evaluation of the effects of centralised cellulosic ethanol and co-products refinery options with sugarcane mill clustering. Biomass Bioenergy 34:1065–1078

Searchinger T, Heimlich R, Houghton RA, Dong F, Elobeid A, Fabiosa J, Tokgoz S, Hayes D, Yu T-H (2008) Use of U.S. cropland for biofuels increased greenhouse gases through land use change. Sci Express 319:1238

Stutter J (2006) LCI of Agricultural production systems, Ecoinvent report No 17

Part III
Impacts on Land Use and Ecosystem Services: Impacts at the National & Regional Scales

Chapter 7
Land Use Change Impacts: National and Regional Scales

Kiyotada Hayashi

7.1 Introduction

Growing demands for food and biofuels are causing deforestation in the tropics. Although the rate of deforestation is decreasing, it is still high and problematic (FAO 2010). Deforestation is mainly the transformation of tropical forest to agricultural land, and it causes environmental problems related to climate change, soil carbon sequestration, ecosystem services, and biodiversity. Reducing tropical deforestation is an international priority especially for the production of Indonesian palm oil and Brazilian soybean oil.

The purpose of this chapter is to perform land use impact assessment within the framework of life cycle assessment (LCA). After observing recent trends in land use change in Indonesia and Brazil, which are important countries in discussing the transformation of tropical forest to agricultural land, land use impact assessment in LCA is reviewed. Plant oils are used as an example to illustrate the importance of land use change in LCA, and the framework for land use impact assessment within LCA is presented. Then, case studies on palm oil production in Indonesia and Malaysia are conducted with illustrating inter-temporal inequality between Europe and Southeast Asia and regionalization of land use impact assessment.

K. Hayashi (✉)
National Agriculture and Food Research Organization, Institute for Agro-Environmental Sciences, 3-1-3 Kannondai, Tsukuba, Ibaraki 305-8604, Japan
e-mail: hayashi@affrc.go.jp

K. Takeuchi et al. (eds.), *Biofuels and Sustainability*, Science for Sustainable Societies, https://doi.org/10.1007/978-4-431-54895-9_7

7.2 Global Land Availability for Biofuels

Recent trends of arable land and forest area in Indonesia and Brazil are shown in Fig. 7.1. Arable land in Indonesia is as a whole increasing, after a small decline in the early 1990s. Arable land in Brazil is also increasing; the increased area during 1990–2008 is more than 10 M ha. In contrast, forest area in both countries is decreasing drastically. 23 M ha forest area (12% of the surface area) in Indonesia and 51 M ha forest area (6% of the surface area) in Brazil have been disappeared during the same period.

These trends illustrate that land availability for biofuels is limited. Global land use for biofuels is estimated at around 13.8 M ha (the sum of the USA, EU, Brazil, and China), which is about 1% of the world cropland (1500 M ha) (Renewable-Fuels-Agency 2008). Large additional land is required for achieving current policy targets of biofuels, even if larger yield increase is possible. Global biofuel targets to 2020 are estimated to require between 56 and 166 M ha (Renewable-Fuels-Agency 2008), which are 1.5 and 4.4 times larger than the surface area of Japan.

7.3 Land Use Change Impacts in LCA

The global land use change discussed so far necessitates the introduction of land use impact assessment into LCA. In this section, after demonstrating the importance of land use change in LCA using the examples of plant oils, the framework of land use impact assessment within LCA is presented.

7.3.1 Importance of Land Use Change in LCA: An Expository Analysis of Plant Oils

The purpose of this expository study is to present a comparative LCA of plant oils at oil mills. Plant oils under investigation are castor oil (India), crude coconut oil (Philippines), jatropha oil (India), olive oil (Cyprus), palm kernel oil (Malaysia), palm oil (Malaysia), palm oil, no clear-cutting (Malaysia), rape oil (Switzerland), rape oil (Europe), rape oil, organic (Europe), soybean oil (Europe), soybean oil (Brazil), and soybean oil (USA).

The system boundaries for plant oil production include agricultural production such as the production of palm fruit bunches and milling processes. Since the former includes production processes of agricultural inputs such as fertilizers and pesticides, the boundaries are termed "cradle to gate." The functional unit is 1 kg of oils. Attributional LCA, which analyzes a single full life cycle, was applied to illustrate typical practices (Reinhard and Zah 2009; Schmidt 2010).

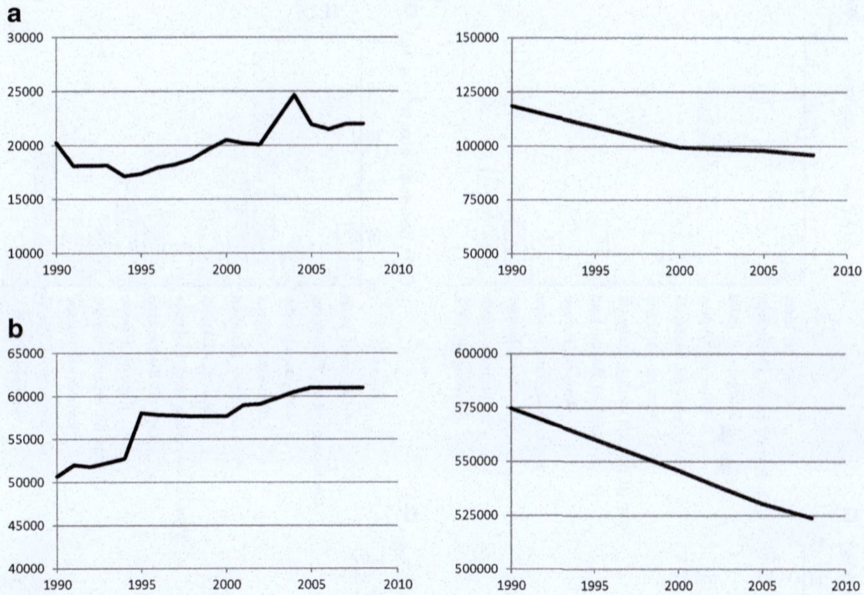

Fig. 7.1 Land use change in Indonesia and Brazil (*Source:* FAOSTAT). (**a**). Arable land in Indonesia (1000 ha), forest area in Indonesia (1000 ha). (**b**) Arable land in Brazil (1000 ha), forest area in Brazil (1000 ha)

A life cycle inventory (LCI) database, ecoinvent 2.2 (Jungbluth et al. 2007), and LCI data prepared by ESU-services (Jungbluth et al. 2009) were used for the analysis. Impact categories used for the assessment are GHG emissions (global warming potential; GWP), energy consumption (cumulative energy demand; CED), and land occupation and transformation (ecosystem damage potential; EDP), which is based on assessment of impacts of land use on species diversity (Koellner and Scholz 2008).

The results of the impact assessment are shown in Fig. 7.2. GHG emissions from plant oil production processes illustrate that the environmental impact of palm oil, especially palm oil without clear-cutting of trees, is relatively small. CED (nonrenewable) demonstrates the same tendency. Since oil palm has the highest oil production efficiency among the oil crops, the values of EDP (occupation) for palm oil are lower than the other oils. However, there is a different trend in EDP (transformation). The value of soybean oil in Brazil is the highest. Although the value of palm oil in Malaysia is high, the value of palm oil without clear-cutting is near zero. In other words, the result is dependent on whether clear-cutting exits or not.

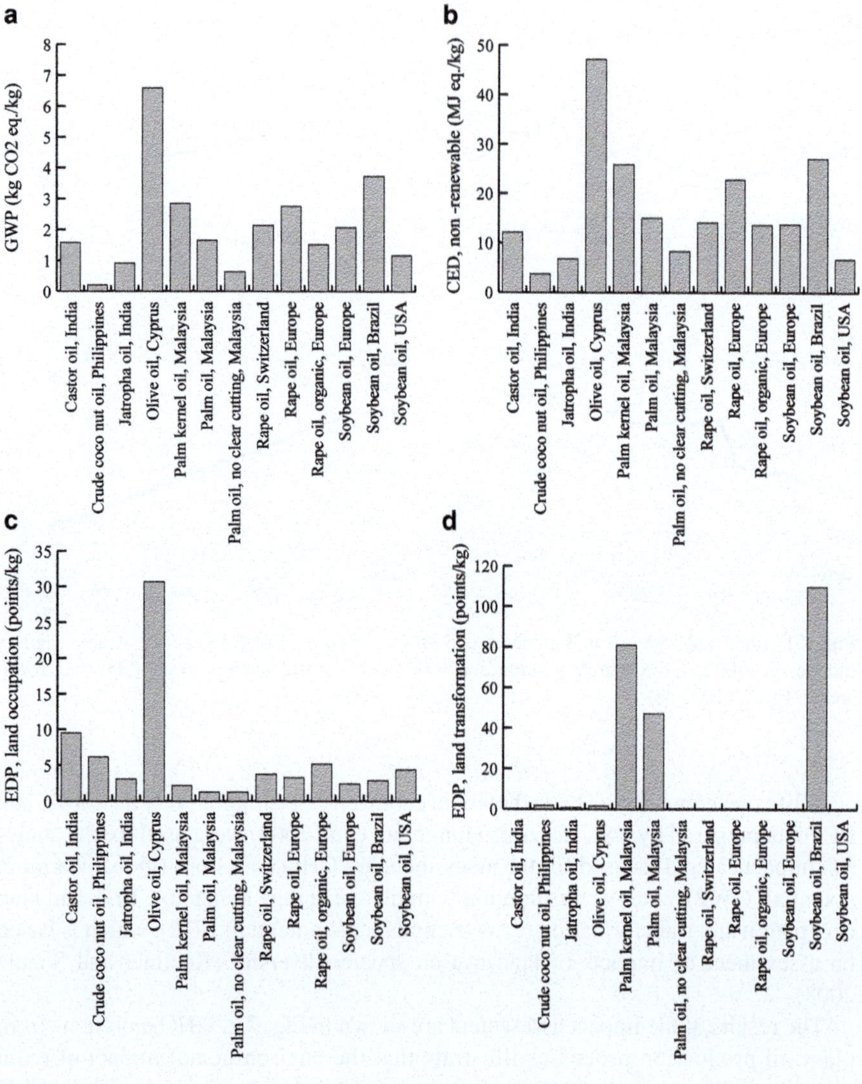

Fig. 7.2 Life cycle GHG emissions, CED, and EDP for various plant oils. (**a**) GHG emissions of plant oils, (**b**) CED for plant oils, (**c**) EDP (occupation) for plant oils, (**d**) EDP (transformation) for plant oils

7.3.2 Methodologies for Land Use Impacts in LCA

One of the striking results in the comparative LCA is the difference between "palm oil, Malaysia" and "palm oil, no clear-cutting, Malaysia." This implies the importance of land use impact assessment in LCA.

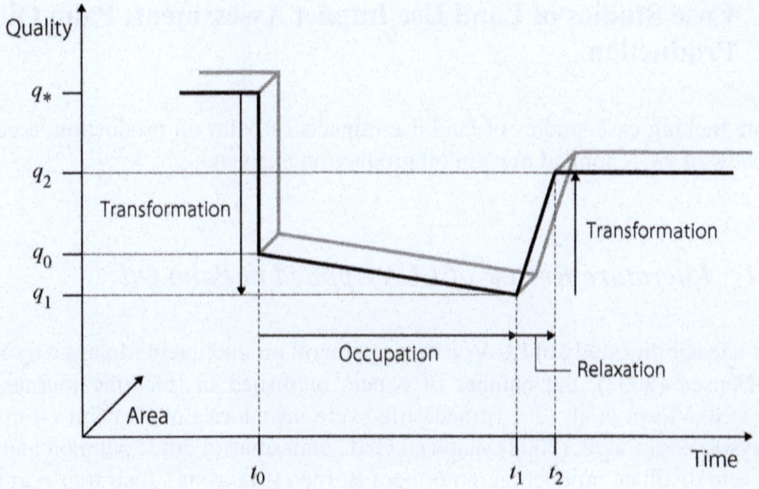

Fig. 7.3 Framework of land use impact assessment in LCA

7.3.2.1 Integration of Land Use into LCA

Land use is an important topic in LCA and gaining much interest in LCA communities recently. Some commodities such as agricultural products have high impacts on land in the production stage, and land use-related environmental impacts should be considered from a product life cycle. The land use impacts are dependent on the location of the land; spatial information will be important. The basic concept in the assessment is ecosystem services of the Millennium Ecosystem Assessment (MEA).

7.3.2.2 Framework for Land Use Impact Assessment Within LCA

The framework for land use impact assessment within LCA is shown in Fig. 7.3, which is based on the UNEP-SETAC task force on the integration of land use impacts into LCA. The characteristics of the framework are summarized as follows: (1) land occupation and land transformation are separated explicitly; (2) land quality is defined during land use (q_*, q_0, q_1, and q_2); (3) the degree of impacts is determined in relation to reference situation (q_*); and (4) the definition of the duration (from t_0 to t_1 or t_2) is important in the assessment.

7.4 Case Studies of Land Use Impact Assessment: Palm Oil Production

Before making case studies of land use impacts of palm oil production, a general overview of LCA applied to palm oil production is given.

7.4.1 *Literature Review of LCA Applied to Palm Oil*

After a feasibility study of LCA on crude palm oil production in Malaysia by Yusoff and Hansen (2007), the number of papers published in scientific journals was increased. Vijaya et al. (2008) made life cycle inventories of 12 palm oil mills in Malaysia. Wicke et al. (2008) analyzed GHG emissions of crude palm oil and palm fatty acid distillate production in northern Borneo (Malaysia), their transport to the Netherlands, and their co-firing with natural gas for electricity production. They stressed that land use change is the most decisive factor in overall GHG emissions and that degraded land should be used for palm oil production. Lam et al. (2009) conducted LCA of palm and jatropha methyl ester (biodiesel) and assessed land requirement, energy balance, and CO_2 emissions and sequestration in Malaysia.

Case studies were not restricted to Malaysia. Angarita et al. (2009) analyzed the life cycle energy balance in palm methyl ester in Brazil and Colombia. Pleanjai and Gheewala (2009) compared the net energy balance and the net energy ratio of palm methyl ester with those of coconut and jatropha methyl ester in Thailand. Papong et al. (2010) analyzed life cycle energy efficiency of palm methyl ester in Thailand.

7.4.2 *Inter-temporal Inequality*

This subsection makes a comparison between plant oil production in Germany and that in Malaysia, to illustrate the relationship between land use impact assessment and policy issues. The inventory data used for both of the oil production are ecoinvent 2.2. The environmental impacts measured are land use impacts on biodiversity developed by Schmidt (2010). Characterization factors for Denmark and Malaysia were used in the assessment, and the unit of the impact is weighted species richness on a standardized area at 100 m^2 (wS_{100}).

The result is depicted in Table 7.1 and summarized as follows: First, the impact of occupation for rape oil is larger than that for palm oil. Second, the impact of transformation from nature to agriculture is larger than that from agriculture to agriculture. Third, the impact of transformation from nature to agriculture in Europe is larger than that in Southeast Asia.

The result implies that there is an inter-temporal inequality between Europe and Southeast Asia. In other words, the environmental impact of past transformation in

Table 7.1 Impact of land use change on species richness (wS_{100}/kg)

Product	Scenario	Occupation	Transformation
Rape oil	Feedstock production: conventional production in Saxony-Anhalt, Germany, from intensive grassland to intensive crop production	0.0191	0.0398
	Oil production: Europe		
Rape oil	Feedstock production: conventional production in Saxony-Anhalt, Germany, from natural forest to intensive crop production	0.0191	6.01
	Oil production: Europe		
Palm oil	Feedstock production: Malaysia, from intensive rubber to agroforestry	0.00473	0.000851
	Oil production: Malaysia		
Palm oil	Feedstock production: Malaysia, from natural forest to agroforestry	0.00473	0.0421
	Oil production: Malaysia		

Europe would be larger than that of current transformation in Southeast Asia; there is a *disadvantage of newcomers*.

7.4.3 Regionalization of Land Use Impact Assessment

Although the previous section clarified the significance of scenarios in land use impact assessment through conducting hypothetical comparisons, more detailed regional conditions have to be specified in land use scenario construction.

7.4.3.1 Regionalization Based on Oil Palm Productivity

The purpose of this subsection is to assess land use impacts of oil palm production. Twelve provinces of Indonesia in Borneo and Sumatra are selected as case study regions (Fig. 7.4). Characterization factors of ecosystem damage potential (EDP) are tentatively used as the impact category.

Inventory data of palm fruit bunches at the farm level (ecoinvent 2.2) were modified to reflect yield differences of oil palm production among provinces. The oil palm productivity data in each province are based on the values of CPO in JIRCAS and MURCI (2006), which are changed into the values of FFB using the conversion coefficient in Corley and Tinker (2003).

The results are shown in Table 7.2. First, EDPs of land transformation are larger than those of land occupation. The former values are in general 38 times larger than the latter values. Second, EDPs of North Sumatra and Riau, the main production areas of oil palm in Indonesia, are relatively low.

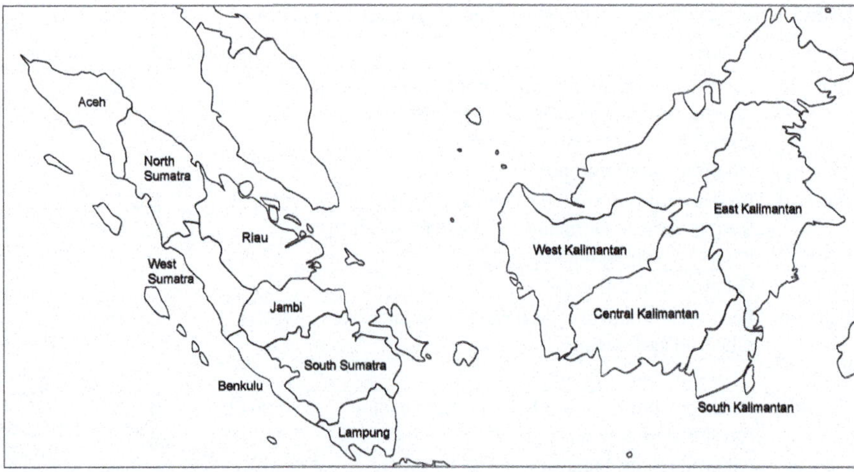

Fig. 7.4 Case study areas in Indonesia

Table 7.2 Ecosystem damage potentials of oil palm production in each region (EDP/kg FFB)

	Occupation	Transformation
Borneo		
West Kalimantan	0.533	20.1
Central Kalimantan	0.518	19.6
South Kalimantan	0.486	18.3
East Kalimantan	0.598	22.6
Sumatra		
Aceh	0.541	20.4
North Sumatra	0.390	14.7
Riau	0.432	16.3
West Sumatra	0.529	20.0
Jambi	0.509	19.2
Bengkulu	0.533	20.1
South Sumatra	0.515	19.4
Lampung	0.551	20.8

7.4.3.2 Regionalization Based on the Share of Peat Land

The above EDP values are useful in understanding the differences of environmental impacts among provinces. However, the conversion of tropical peat land is not considered in the assessment. Thus, this subsection introduces the share of peat land in each province into the assessment. CO_2 emissions from land transformation are used as the criteria.

Table 7.3 CO_2 emissions from land transformation to oil palm production in each region (kg CO_2/kg FFB)

	CO_2 emissions from land transformation			
	Case 1	Case 2	Case 1/Case 2	Fossil CO_2 emissions
Borneo				
West Kalimantan	0.861	0.289	3.0	0.165
Central Kalimantan	0.491	0.281	1.8	0.164
South Kalimantan	0.263	0.263	1.0	0.162
East Kalimantan	0.326	0.324	1.0	0.170
Sumatra				
Aceh	0.705	0.293	2.4	0.166
North Sumatra	0.656	0.211	3.1	0.156
Riau	1.038	0.234	4.4	0.159
West Sumatra	1.417	0.287	4.9	0.165
Jambi	0.824	0.276	3.0	0.164
Bengkulu	1.061	0.289	3.7	0.165
South Sumatra	0.560	0.279	2.0	0.164
Lampung	0.712	0.298	2.4	0.167

Case 1: The percentage of peat soil is considered
Case 2: The percentage of peat soil is not considered

In the calculation process, land transformation is separated into two parts: one is the provision of stubbed land (clear-cutting of primary forest) and the other is the production of palm fruit bunches at the farm level (direct emission). The former is the transformation from tropical rain forest to intensive forest by clear-cutting, and the latter is the transformation from intensive forest to intensive short rotation forest. Burning of the 20% of the biomass is supposed in clear-cutting. CO_2 emissions from land transformation are counted in direct emissions from the stubbed land provision process and direct emissions from the oil palm production process. The percentages of peat soil in oil palm plantations in each province are taken from Koh et al. (2011). CO_2 emissions from peat decomposition related to oil palm plantations are based on Uryu et al. (2008).

The results are summarized as follows (Table 7.3): First, CO_2 emissions from land transformation are larger than fossil CO_2 emissions even if peat land is not considered. Second, if we introduce the percentages of peat land into the calculation, the values increase drastically. The values for West Sumatra and Riau increase more than four times and for Bengkulu, North Sumatra, Jambi, and West Kalimantan increase more than three times. CO_2 emissions from land transformation with considering peat land in West Kalimantan are more than eight times larger than fossil CO_2 emissions. These results indicate the importance of regional land conditions in the assessment.

7.5 Discussions

From the land use trends at the global level, the focus of attention in this chapter has evolved into the province level. Since the next analytical step will be the detailed assessment at the plantation level, the first point to be discussed here is LCI data. It is important to point out that the background data used for modeling agricultural production such as fertilizers and pesticides in ecoinvent are European or Swiss ones. Therefore, adaptation and development of the regionalized data are important; e.g., Indonesian background data are appropriate for LCA of palm oil in Indonesia.

A pragmatic method to regionalize background data is the modification of European LCI data. In other words, if local industrial LCI data (as background data for, e.g., fertilizer production) exist, they can be used to regionalize the European LCI data (e.g., LCIs of fertilizers). Ossés de Eicker et al. (2010) demonstrate the usefulness of the modification method through conducting a case study of triple superphosphate in Brazil. Hayashi et al. (2010) illustrate the effectiveness of the method in constructing LCI database for crop production systems in Japan. Recent LCI inventories of palm oil production developed by, for example, Malaysian Palm Oil Council (MPOC) and Indonesian Oil Palm Research Institute (IOPRI) will play an important role in conducting LCA of palm oil.

The second point is how to cope with indirect impacts. Indeed, there are intensive discussions on indirect land use change after the *Science* papers (Fargione et al. 2008; Searchinger et al. 2008). Although the assessments in this chapter are restricted to direct impacts because there is no consensus in the methodology including consequential modeling, indirect effects of biofuels should not be neglected (Renewable-Fuels-Agency 2008; Miller et al. 2010).

7.6 Concluding Remarks

The implications of the case studies in this chapter are summarized as follows: First, land use impact assessment within the framework of LCA can be applied to intertemporal comparisons based on hypothetical scenario construction. The result revealed that there is a disadvantage of newcomers and it means social issues are deeply related to environmental impact assessment of land use change. Second, land use impact assessment can be regionalized using land productivity and conditions. The province-level assessment showed that land transformation is crucial both in reducing GHG emissions and in conserving biodiversity. Although land use impact assessment within LCA is still on the development, it will play an important role in policy-making processes.

References

Angarita EEY, Lora EES, da Costa RE, Torres EA (2009) The energy balance in the Palm Oil-Derived Methyl Ester (PME) life cycle for the cases in Brazil and Colombia. Renew Energy 34(12):2905–2913. doi:10.1016/j.renene.2009.05.007

Corley RHV, Tinker PB (2003) The oil palm, World Agriculture Series, vol 9, 4th edn. Wiley-Blackwell, Oxford

FAO (2010) Global forest resources assessment 2010: main report. Food and Agriculture Organization of the United Nations, Rome

Fargione J, Hill J, Tilman D, Polasky S, Hawthorne P (2008) Land clearing and the biofuel carbon debt. Science 319(5867):1235–1238. doi:10.1126/science.1152747

Hayashi K, Uchida S, Hokazono S, Sato M (2010) Modeling life cycle inventories for crop production in Japan: development of the NARO LCI database. Paper read at LCA Food 2010: VII International Conference on Life Cycle Assessment in the Agri-Food Sector, at Bari, Italy

JIRCAS and MURCI (2006) Final report for study of biomass production and flow in Indonesia (Phase II). Japan International Research Center for Agricultural Science and MU Research and Consulting Indonesia

Jungbluth N, Chudacoff M, Dauriat A, Dinkel F, Doka G, Faist Emmenegger M, Gnansounou E, Kljun N, Schleiss K, Spielmann M, Stettler C, Sutter J (2007) Life cycle inventories of bioenergy: ecoinvent report no. 17, ecoinvent report. Swiss Centre for Life Cycle Inventories, Dübendorf

Jungbluth N, Büsser S, Stucki M, Leuenberger M (2009) Life cycle inventories of food consumption: EcoSpold LCI database of ESU-services. ESU-services Ltd. Available from http://www.esu-services.ch/inventories.htm

Koellner T, Scholz R (2008) Assessment of land use impacts on the natural environment. Int J Life Cycle Assess 13(1):32–48. doi:10.1065/lca2006.12.292.2

Koh LP, Miettinen J, Liew SC, Ghazoul J (2011) Remotely sensed evidence of tropical peatland conversion to oil palm. Proc Natl Acad Sci 108(12):5127–5132. doi:10.1073/pnas.1018776108

Lam MK, Lee KT, Mohamed AR (2009) Life cycle assessment for the production of biodiesel: a case study in Malaysia for palm oil versus jatropha oil. Biofuels Bioprod Bioref-Biofpr 3(6):601–612. doi:10.1002/bbb.182

Miller FP, Vandome AF, McBrewster J (2010). Indirect land use change impacts of biofuels. Alphascript Publishing

Ossés de Eicker M, Hischier R, Hurni H, Zah R (2010) Using non-local databases for the environmental assessment of industrial activities: the case of Latin America. Environ Impact Assess Rev 30(3):145–157. doi:10.1016/j.eiar.2009.08.003

Papong S, Chom-In T, Noksa-nga S, Malakul P (2010) Life cycle energy efficiency and potentials of biodiesel production from palm oil in Thailand. Energ Policy 38(1):226–233. doi:10.1016/j.enpol.2009.09.009

Pleanjai S, Gheewala SH (2009) Full chain energy analysis of biodiesel production from palm oil in Thailand. Appl Energy 86:S209–S214. doi:10.1016/j.apenergy.2009.05.013

Reinhard J, Zah R (2009) Global environmental consequences of increased biodiesel consumption in Switzerland: consequential life cycle assessment. J Clean Prod 17:S46–S56. doi:10.1016/j.jclepro.2009.05.003

Renewable-Fuels-Agency (2008) The Gallagher review of the indirect effects of biofuels production. Renewable Fuels Agency, St Leonardo-on-Sea

Schmidt JH (2010) Comparative life cycle assessment of rapeseed oil and palm oil. Int J Life Cycle Assess 15(2):183–197. doi:10.1007/s11367-009-0142-0

Searchinger T, Heimlich R, Houghton RA, Dong F, Elobeid A, Fabiosa J, Tokgoz S, Hayes D, Yu T-H (2008) Use of U.S. croplands for biofuels increases greenhouse gases through emissions from land-use change. Science 319(5867):1238–1240. doi:10.1126/science.1151861

Uryu Y et al. (2008) Deforestation, forest degradation, biodiversity loss and CO2 emissions in Riau, Sumatra, Indonesia. In: WWF Indonesia technical report. Jakarta

Vijaya S, Ma AN, Choo YM, Nik MNS (2008) Life cycle inventory of the production of crude palm oil – a gate to gate case study of 12 palm oil mills. J Oil Palm Res 20:484–494

Wicke B, Dornburg V, Junginger M, Faaij A (2008) Different palm oil production systems for energy purposes and their greenhouse gas implications. Biomass Bioenergy 32(12):1322–1337. doi:10.1016/j.biombioe.2008.04.001

Yusoff S, Hansen SB (2007) Feasibility study of performing an life cycle assessment on crude palm oil production in Malaysia. Int J Life Cycle Assess 12(1):50–58. doi:10.1065/lca2005.08.226

Chapter 8
Socioeconomic Impacts of Biofuels in East Asia

Mark Elder, Jane Romero, Anindya Bhattacharya, Daisuke Sano, Naoko Matsumoto, and Shinano Hayashi

8.1 Introduction

This chapter discusses the social and economic impacts of biofuels in East Asia by analyzing four country case studies. Three case study countries are large Asian rapidly developing countries which were expected to be large consumers and producers of biofuels at the beginning of the biofuel boom in the late 2000s: Indonesia, India, and China. All three of these countries developed ambitious initial biofuel promotion plans. The fourth country case is a developed country, Japan. Japan has some domestic production potential, although it is quite small compared to potential domestic demand, so many expected that Japan might become a significant importer of biofuels or biofuel feedstocks, especially from the Asian region.

The main potential positive impacts for all four countries include employment, income, rural development, and energy security. Rural electrification and increasing energy access for poor people are important objectives for developing countries. Air pollution reduction is another potential benefit, although this varies by the type of fuel and feedstock. The main potential negative impacts include competition with food and other land uses; negative impacts on ecosystem services, particularly related to deforestation and water usage; and social impacts such as land tenure rights (e.g., if land of poor farmers is taken over by large producers without consent or fair compensation).

Several important factors should be taken into account when analyzing impacts. First, the potential effects vary significantly by feedstock, market structure and conditions, and other local conditions such as geography, social structure, etc. Second, there may be difficult trade-offs between economic costs and desired socioeconomic impacts. For example, maximizing employment and income for farmers and

M. Elder (✉) · J. Romero · A. Bhattacharya · D. Sano · N. Matsumoto · S. Hayashi
Institute for Global Environmental Strategies (IGES),
2108-11 Kamiyamaguchi, Hayama, Kanagawa 240-0115, Japan
e-mail: elder@iges.or.jp

K. Takeuchi et al. (eds.), *Biofuels and Sustainability*, Science for Sustainable Societies, https://doi.org/10.1007/978-4-431-54895-9_8

workers may require labor-intensive, smaller-scale production methods with higher wages. In contrast, biofuel producers will generally prefer large-scale production methods to minimize costs and maximize profits, including labor-saving technology. Moreover, large-scale production, which is generally more cost-efficient and profitable, may result in large-scale deforestation and significant negative effects on ecosystem services. Third, impacts (both positive and negative) may be shifted to other countries if biofuels and/or feedstocks need to be imported or if domestic production of biofuels displaces other domestically produced goods and services. Fourth, measurement of impacts is often difficult and hampered by a lack of data.

The rest of this chapter surveys the four country cases. Each case includes an overview of each country's main biofuel-related policies and market conditions, discussion of the main socioeconomic impacts, and consideration of the perspectives of different stakeholders. The chapter concludes with a comparison, synthesis, and a discussion of the policy implications.

8.2 Indonesia

8.2.1 Overview of Indonesia's Main Policies

Indonesia's energy policy has been focused on the goals of energy security and promoting access to energy in the face of sharply rising energy consumption due to rapid economic growth. It used to be an OPEC member with a significant oil surplus, but it became a net importer in 2004. Indonesia has subsidized fossil fuels for transport and cooking heavily since 1967. By 2005, the burden of these subsidies became very high as the government spent more than $8 billion to subsidize the market price of petroleum fuels (IEA 2008). Facing declining oil reserves and mounting subsidies, the government enacted Presidential Decree No.5/2006, the so-called Mixed Energy Policy, to diversify Indonesia's energy sources to include renewable energy and biofuels. The transport sector uses at least 30% of liquid fuels in Indonesia. Electricity access in rural areas is low with over 70 million Indonesians estimated to be unconnected to power grids (Jayawardena 2005). The potential of biofuels as a transport fuel substitute, source of fuel in rural areas, and low agricultural commodity prices at that time motivated the government to pursue biofuel development. The export potential of biofuels also appeared to be highly lucrative as Annex 1 countries sought cleaner fuel alternatives to meet their Kyoto Protocol carbon emission reduction targets.

Presidential Instruction No.1/2006 aimed to accelerate biofuel utilization as a fossil fuel substitute. Presidential Regulation No.5/2006 on National Energy Policy expected the share of oil in national energy consumption to be reduced to 20% by 2025, while the share of biofuels should increase to at least 5% in the national energy mix as shown in Fig. 8.1.

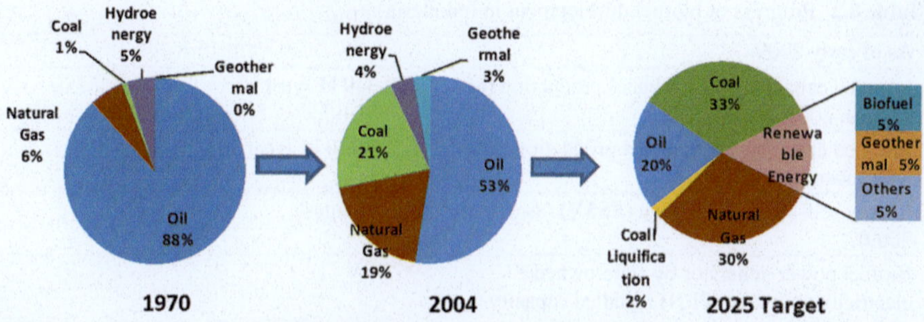

Fig. 8.1 Energy mix trends and targets in Indonesia

Table 8.1 Indonesia's roadmap for biofuel development

Biofuel Type	Unit	2005–2010	2011–2015	2016–2025
Biodiesel	Percent consumption (of diesel fuel)	10%	15%	20%
	Amount (million kL)	2.41	4.52	10.22
Bioethanol	Percent consumption (of gasoline)	5%	10%	15%
	Amount (million kL)	1.48	2.78	6.28
Bio-oil/bio-kerosene	Amount (million kL)	1	1.8	4.07
Bio-oil/pure plantation oil (PPO)	Amount (million kL)	0.4	0.74	1.69
Biofuel	Percent consumption (of energy mix)	2%	3%	5%
	Amount (million kL)	5.29	9.84	22.26

Source: Legowo 2009

Presidential Decree No. 10/2006 established the National Team for Biofuel Development for poverty and unemployment alleviation which was mandated to draft the national blueprint for biofuel development. The road map of biofuel development (refer to Table 8.1) in Indonesia identified crude palm oil (CPO) and *Jatropha curcas* as the main feed stocks for biodiesel, and sugarcane and cassava as the main feed stocks for bioethanol (Kusdiana 2006). The Indonesian government set blending mandates at 10% for biodiesel effective from 2010 and 20% for bioethanol starting in 2015 with the target of producing 17.3 billion liters of bioethanol and 29 billion liters of biodiesel by 2025. To kick start the program, the government instructed the national oil company, Pertamina, to start selling biodiesel with a 5% blend produced from palm oil.

According to the plan, biofuel development was expected to enhance the rural economy, job creation, and poverty alleviation. The plan expected that 3.5 million jobs would be created by 2010, which could increase up to 6.9 million jobs in 2025. In the long run, the generation of energy from locally available renewable sources through the Energy Self-sufficient Village (ESSV) program and the Special Biofuel

Table 8.2 Progress of biofuel development in Indonesia

As of early 2009	
Installed capacity for bioethanol production (as of June, 2008)	192,349 kL/year
Installed capacity for biodiesel production (as of December, 2008)	2,529,110 kL/year
Energy self-sufficient village (ESSV) (as of early 2009)	150 villages
Biofuel power generator by state-owned electricity company (PLN) installed capacity (as of June, 2008)	96 MW
Biofuel utilization in industry (as of November, 2008)	5%
Biofuel utilization in electric sector by PLN (PLTGU Gresik 19 MW started in 31 January 2009)	8 kL/day
Projection:	
Bioethanol development: Projection up to 2010	~4,000,000 kL/year
Biodiesel production: Projection up to 2010	~5,000,000 kL/year
Biofuel power generator by PLN in 2009–2010	220,000 kL biodiesel in all biofuel power generators allover Indonesia

Source: Legowo 2009

Zone (SBZ) program by encouraging each region to develop its biofuel potential were expected to contribute to national energy security.

The progress of the implementation of the biofuel development plan in Indonesia as of early 2009 is illustrated in Table 8.2 above.

8.2.2 Overview of Main Biofuel Market Conditions in Indonesia

8.2.2.1 Biodiesel from Palm Oil and *Jatropha*

For biodiesel, palm oil is the main feedstock, based on Indonesia's well-established palm oil industry, with a total plantation area estimated to be over 6 million hectares. Indonesia surpassed Malaysia to become the world's largest palm oil producer in 2008, producing 18 billion liters (OECD-FAO 2008).

The global price of palm oil soared from mid-2006 to the middle of 2008, partly due to its popularity as a major biodiesel feedstock. As a result, biofuel production from palm oil became unprofitable. Initial estimates that palm oil-based biodiesel would be competitive to conventional oil at $400 per metric ton, or about $54 per barrel, proved to be wrong. When oil prices peaked above $140 per barrel, the price of palm oil rose even higher making biofuels more expensive to produce. Pertamina suffered losses from its biofuel blends because the government required it to sell

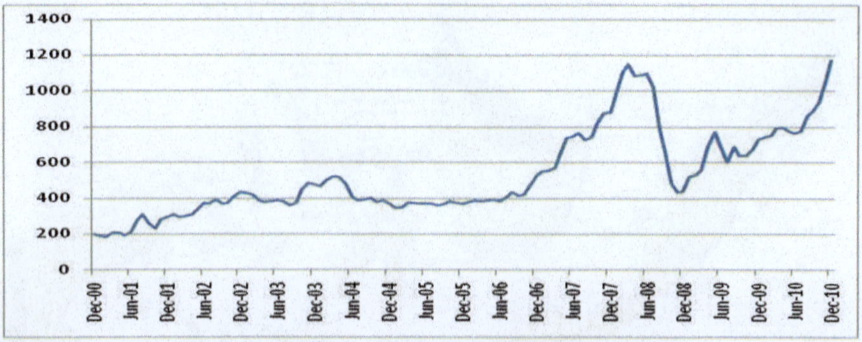

Source: Commodity Price Indices. http://www.indexmundi.com/

Fig. 8.2 Palm oil monthly prices in US dollars per metric ton

biofuels at the same price as subsidized petroleum but did not provide additional subsidies to cover the higher costs of biofuel production (GSI 2008a, b). Therefore, Pertamina reduced the biodiesel content to barely 1%.

Palm oil is also very important for cooking in Indonesia, so the government became very concerned about surging prices. In response, export taxes were imposed on crude palm oil to discourage exports and prioritize its use for cooking, and the government also imposed a 2% export tax on biofuels (Leow 2008; Commodity Online 2008).

The government also became concerned about the contentious debate on the environmental impacts of biofuels – especially relating to the conversion of forests to biofuel feedstock monoculture plantations – and concerns about their role in raising food prices also grew worldwide (e.g., Fargione et al. 2008; Searchinger et al. 2008; Pimentel et al. 2007). Land use change and deforestation in Indonesia were identified as so significant that the country ranked third in total GHG emissions globally. In 2008, the EU reviewed its biofuel mandate and stopped importing oil palm from Indonesia and Malaysia citing environmental concerns (USAID-Asia 2009). The high price of palm oil and questions about its sustainability had a significant impact, and many refineries stopped operations and stalled plans for expansion and new development. When palm oil prices declined in the later part of 2008, biofuel production levels increased once again, but this was short-lived as prices increased again as shown in Fig. 8.2 (Reuters 2007; GSI 2008).

The government also promoted *Jatropha* as a biodiesel feedstock, recognizing the volatility of palm oil prices, to avoid the food–fuel conflict. The biofuel roadmap initially set a target of 1.5 million hectares of previously logged and nonproductive land to be planted with *Jatropha*, as shown in Fig. 8.3, but as of 2008, only 10% was planted. Initial demand for *Jatropha* seeds to make seedlings significantly raised their price, generating interest from many investors and farmers. When demand for seeds stabilized, actual yield was low (only about one-fourth of the initial estimates of at least 5 tons per hectare per year), making it unprofitable to process them for biodiesel. Research to create high-yielding varieties has continued.

Fig. 8.3 Target areas for *Jatropha* plantation in Indonesia

8.2.2.2 Bioethanol from Sugarcane and Cassava

Fuel ethanol in Indonesia is produced from sugarcane molasses. Total ethanol production in Indonesia was about 212 million liters in 2008 (OECD-FAO 2008), produced by four fuel ethanol plants operating with a combined capacity of 14 million liters per year (GSI 2008b). To comply with the initial 10% blending mandate in 2010, the goal was to produce nearly 4 billion liters (APEC 2008). The mandatory blending ratio was reduced to 3% in 2010.

There may be some scope to increase the efficiency of sugarcane production. About 2 million hectares of land is used for sugarcane production in Indonesia. However, 50% of sugarcane producers are small holders, and the average farm size is barely half a hectare, so there is room to increase the scale of farms. In addition, there are many small sugar mills which still use outdated technology. The government also considered cassava as alternative ethanol feedstock. In 2006, about 650,000 hectares were planted with cassava. As in the case of sugarcane, producers are mostly small holders producing cassava chips, while the large processors produce starch. A high-yield variety was initially introduced in Java to improve the current harvest yield of 15–18 tons/ha. Only 0.5% of cassava is used for bioethanol as it is mainly used for direct consumption and food processing. In addition, there are not many fully functional bioethanol plants utilizing cassava yet. The situation may change, and more cassava could be used for bioethanol production if bioethanol producers would offer a higher price to farmers than the food processing industry. MEDCO inaugurated Indonesia's first bioethanol plant utilizing cassava as feedstock in Lampung in late 2009, and it is now operating at full scale. They pay a premium to ensure availability of cassava to maintain their operations.

8.2.3 Socioeconomic Impacts

Biofuel production in Indonesia, regardless of feedstocks used, is not yet economically viable under current conditions and requires heavy subsidies. Economic sustainability is essential for long-term biofuel development plans like Indonesia's, especially in light of the country's increasing fiscal constraints. In order to justify support from the government budget, the social benefits of biofuels need to be demonstrated. In this context, the government also considered how to use biofuels to promote rural development.

The government launched the Energy Self-sufficient Village (ESSV) program in 2006 targeting 1000 villages in remote areas to be self-sufficient in their energy needs by utilizing their own local renewable energy resources. Of the 1000 villages, 500 will produce their own supply of biofuels from *Jatropha*, cassava, or sweet sorghum to run basic equipment for lighting and farm activities and to replace the use of kerosene for cooking purposes. The other 500 villages will harness their water resources to develop mini-hydro or pico hydropower and install solar photovoltaics (PV). As of 2010 the biofuel-based project was implemented in almost 150 villages.

For this study, three pilot ESSV projects were visited and surveyed – Karangtengah village in Wonogiri utilizing cassava for bio-kerosene production, Purwantono village in Wonogiri utilizing sweet sorghum for bio-kerosene production, and Way Isem village in Lampung utilizing *Jatropha* for biodiesel production. The farmers in Karangtengah have sufficient experience in planting cassava which they sell for processing as food or animal feed. The village allocated some common land to increase cassava plantation to be used for bio-kerosene production. At the time of the survey, the mini-processing plant had been constructed but was still undergoing intermittent testing to achieve consistency in the desired blend (~70% ethanol). The potential to produce bio-kerosene out of cassava was welcomed eagerly by farmers, but the project remained a community experiment, and how it would be managed and sustained remained to be seen. The case in Purwantono was more complicated because the farmers had no prior knowledge in planting sweet sorghum. The mini-processing plant had also been built, but even the necessary testing was difficult to conduct because of a lack of feedstock. In Way Isem, farmers planted *Jatropha* as hedges and in idle plots. Initially there was high demand for seeds and seedlings, so many farmers planted *Jatropha*. However, farmers did not want to become full-time *Jatropha* farmers because they earned more from planting other crops. From the limited operations in Way Isem, and also because of a lack of *Jatropha* seeds, farmers valued more the *Jatropha* waste that could be used to produce biogas for cooking than the straight *Jatropha* oil.

Overall, the success rate of ESSV was lower than expected despite the government's assistance providing the necessary processing equipment. The feedstock supply was too unstable to operate continuously. Coordination among agencies involved in the implementation was weak. What the farmers in the villages primarily needed was the know-how to improve production yields either by having access to high-yield varieties or by improving their farming practices sufficiently for them to be encouraged to plant the required biofuel feedstocks for energy purposes. However, government funding was mostly allocated for procuring equipment and building mini-processing facilities. To ensure a stable supply of feedstocks, farmers should be assisted to improve their productivity and there should be more efforts to help farmers understand the agricultural and energy benefits of biofuels (Romero 2010).

8.2.4 Analysis

In 2006, Indonesia drafted a comprehensive national biofuel development policy with the dream of becoming the "Middle East of biofuels." The policy was undermined even before it was fully implemented by the events leading to the sharp rise and fall of oil prices in 2008. By 2009, the government, industry, civic organizations, and farmer groups were reconsidering their euphoric expectations for biofuels. The government's flexible response to reduce blending targets was laudable, as rigidly adhering to the initial targets likely would have meant more losses.

In hindsight, the policies assumed that the groundwork for establishing the biofuel industry had already been laid. Initially, most policy discussions focused on trade and investment, neglecting the vital role of the agriculture sector. Moreover, important assumptions underlying the expectations of the economic viability of biofuel projects were proved to be incorrect. In the case of palm oil, it was assumed that the palm oil price would be lower than the oil price. Farmers gained when the price of palm oil went up, although the nascent biodiesel industry nearly collapsed. For *Jatropha*, the actual yield of *Jatropha* seeds was only about one-fourth of what had been projected, but the necessary agricultural inputs were more than initially estimated. Small holders were the ones most adversely affected since they did not have much capital to offset their losses.

Overall, Indonesia still has the potential to build a flourishing biofuel industry. To achieve it, lessons learned should be incorporated in rethinking the national biofuel policy. Action plans to complement the national policies should be included, as the lack of action plans caused confusion and competition instead of coordination among relevant agencies. Capacity training and R&D measures should be strengthened. And the most critical of all is to eliminate fossil fuel subsidies and shift the funds to support cleaner energy sources.

8.3 India

8.3.1 Overview of India's National Policies on Biofuels

Indian national biofuel policy, in 2009, was cautiously optimistic in nature. It aimed to achieve a 20% blending of biofuels with gasoline by 2017, mainly from ethanol. However, the 10% ethanol blending target set in October, 2008, was not achieved in the country, and the 20% target seemed quite challenging. In the national policy, ethanol was envisaged as the major source of biofuels in the country, while the other plant-based biofuels (mainly biodiesels) were considered as secondary sources.

The major obstacle to maintaining a stable supply of ethanol for biofuel production was the instability of sugarcane production. The pricing of ethanol-based biofuels was also very controversial, and disagreements between the sugar industry, the major producer of ethanol (from its by-product molasses), and the oil marketing companies, the main distributors of the biofuels, were not resolved. National policy briefly mentioned pricing, but the government gave no clear indication regarding how it will handle the issue except to pass the responsibility to the Biofuel Steering Committee. Uncertainty about the pricing policy was a serious obstacle to the promotion of the bioethanol industry in India.

Regarding biodiesel production, the national policy stated that no food-producing land should be used, and biodiesel should be produced only from nonedible oilseed plantations on lands which were considered wastelands, degraded, or marginal. However, it was not clear just how much wasteland was available. Land availability is a serious problem in India where food shortages are increasing. Definitions of degraded and wasteland vary according to productivity and length of fallowness. Agricultural experts in the country claimed that technology is available to convert a majority of the so-called wasteland to at least mono-cropping land provided required inputs are given. Moreover, most of the degraded lands are either forest lands, which are difficult for farmers to access, or village common lands belong to the panchayats and communities which are used by the landless and tribal communities for cattle grazing and other purposes. The Ministry of New and Renewable Energy, the implementing agency of the national biofuel policy, has little ability to procure wasteland to produce biodiesel because land is under the jurisdiction of other ministries and departments. The land may seem to be "wasted" and "barren" to outsiders, but in reality much of it provides sustenance for millions of poor and marginalized rural people. Most of these wastelands are classified as common property resources (CPRs) and are used as grazing ground for the village cattle. So on one hand, ethanol supply fluctuates, and on the other hand, availability of wasteland for nonedible oil seed production is also uncertain under India's new national policy. Therefore, two of the pillars of biofuel policies (ethanol and wasteland) have been uncertain for India. Nevertheless, the policy also ensured the use of the National Employment Guarantee Act (NREGA) to provide financial support for the labor costs of biofuel production. Unfortunately, NREGA was the only source of funds for rural employment available for many government activities within the village areas (not only biofuel policy), so there was a shortage of funds for biofuel activities.

Apart from the central government's policy on biofuels, there was also a variety of initiatives by state governments, mainly with private sector partnerships. For example, the state of Andhra Pradesh entered into a formal agreement with Reliance Industries to plant *Jatropha* on 200 acres (0.81 km2) of land at Kakinada for high-quality biodiesel. The state of Chhattisgarh decided to plant 160 million saplings of *Jatropha* in all of its 16 districts with the aim of becoming a biofuel self-sufficient state by 2015, and it planned to earn Rs. 40 billion annually after 2010 by selling seeds. In September 2007, the Hindustan Petroleum Corporation Limited (HPCL) and the Maharashtra State Farming Corporation Ltd. (MSFCL) created a *Jatropha* seed-based biodiesel joint venture with a 500 acre *Jatropha* plantation in the so-called degraded forest areas of the state. Indian Railways has started using biodiesel from *Jatropha* for its diesel engines. However, despite these initiatives, no commercial production of biodiesel on a national scale has been recorded.

8.3.2 *Status of the Indian Biofuel Market*

With the gradual increase in demand for renewable energy, the biofuel sector in India has taken the necessary steps toward large-scale commercial production of fuel crops. With the primary objective of increasing the production of biofuels, namely, biodiesel and bioethanol, the Government of India took the lead and formulated the National Mission on Biodiesel in July 2002. In order to avoid creating a food–fuel conflict in the country, the government from the beginning encouraged the use of fermented sugarcane molasses and nonedible oil seeds. So, in India ethanol is produced through fermentation of sugarcane molasses, and biodiesel is produced through transesterification of nonedible oils from *Jatropha curcas*, pongamia, neem, etc.

8.3.2.1 Bioethanol from Sugarcane

Due to robust economic growth in India, transport fuel demand has also increased at a very high rate. Moreover, demand for ethanol has also increased to meet the blending target of 5% of total transport fuel set by the National Government in 2003. Table 8.3 shows the projected demand and supply of ethanol in the Indian market. This clearly indicates that 5% blending seems feasible but 20% blending only from ethanol may be quite difficult and unrealistic.

In order to meet the ethanol-based biofuel target, it would be necessary for India to maintain a steady production level of sugarcane over the target's time period. However, the yield of sugarcane in India varies from an average of 77 tons/ha in tropical states to about 52 tons/ha in subtropical states, and it also varies under

Table 8.3 Projected demand and supply of ethanol for 5% blending in petrol

Year	Petrol demand (Mt)	Ethanol demand (M L)	Molasses prodn. (Mt)	Ethanol production (M L)			Ethanol utilization (M L)		
				Molasses	Cane	Total	Potable	Industry	Balance
2001–2002	7.07	416.14	8.77	1775	0	1775	648	600	527
2006–2007	10.07	592.72	11.36	2300	1485	3785	765	711	2309
2011–2012	12.85	756.36	11.36	2300	1485	3785	887	844	2054
2016–2017	16.4	965.3	11.36	2300	1485	3785	1028	1003	1754

Source: Planning Commission (2003)
The above information is based on the following assumptions:
a-1. The area under cane cultivation is expected to increase from 4.36 Mha in 2001–2002 to 4.96 Mha in 2006–2007 which would result in an additional cane production of 50 MT.
a-2. About 30% of cane goes for making gur (jaggery) and khandsari (unrefined sugar). If there is no additional increase in khandsari demand, sugar and molasses production would increase.
a-3. The present distiller capacity is for 2,900 million liters (M L) and appears to be sufficient for 5% blend until 2016–2017.
a-4. Annual demand growth of 3% for potable ethanol and 3.5% for industrial ethanol.

different irrigation conditions. The average yield of sugar is approximately 105 kg per ton of cane, and about 40 kg of molasses is produced per ton of cane from which about 10 l of ethanol can be obtained. If the sugarcane is directly and fully used in ethanol production, the yield of ethanol is 70 l per ton (Gonsalves 2006).

The production cost of ethanol in India in 2009 was between Rs.14 and 20 per liter, depending upon the source, which is still comparable to the market price of gasoline. However, this cost is before tax. After sales, excise, and other direct and indirect taxes, the ethanol price is as high as other fuels and may need a selling price subsidy to compete with the standard fuels to meet the 5% blending target. It has been observed that the major reason for the high production cost of ethanol is the increasing cost of sugarcane production in India. Unfortunately, the cost of sugarcane production in India is expected to continuously increase mainly due to a shortage of water resources and its impact on reduced productivity, continued use of low-quality sugarcane species, unscientific sugarcane cultivation methods, and lack of a market-based pricing mechanism for sugar. In addition, increasing the efficiency of sugarcane processing, including juice extraction and fermentation, is also important to bring down the final cost of ethanol production. Finally, although sugarcane-based ethanol is commercially a viable option for India to produce biofuels, the increasing costs of producing ethanol are a serious threat to its economic viability in the long run.

8.3.2.2 Biodiesel From Nonedible Oilseeds

According to the national policy on biofuels, plant-based nonedible oilseeds were expected to supply biodiesel along with bioethanol together to meet the national target of 20%. It was envisaged that around 400 different types of nonedible oilseeds are available in India which could produce the necessary amount of biodiesel. However, biodiesel in India has been virtually a nonstarter. *Jatropha* is one of the major feedstocks for biodiesel production in India, but unfortunately it has performed poorly. The reasons for this include the following technical problems and policy deficiencies:

- There is a lack of infrastructure for seed collection and oil extraction. In the absence of infrastructure and available oilseeds, it will be difficult to persuade entrepreneurs to invest in transesterification plants. Collection of nonedible oil seeds is a manual operation, and for a large biodiesel plant, it is a logistical nightmare. In 1 day, a person can collect up to 80 kg of seeds, which can produce 20–23 l of oil. The collection is done for 3 months, once or twice a year. For a plant with a capacity of 100 tons per day (8 million gallons per year), 15,000 people are necessary to collect the seeds. Organizing such a large part-time labor force is a major challenge.
- The *Jatropha* plant takes 24–30 months to flower and produce seeds. To promote widespread *Jatropha* farming, the livelihood of the farmers in the intervening period, without an income from the *Jatropha* crop, must be secured. At the time of this research there was no way to achieve this in the market except for privately funded projects. In particular, this is a problem for landless farmers and laborers who do not qualify for any interim payments since they do not own the land.
- There have been some uncertainties about how much inputs (irrigation, fertilizers) are needed to realize commercially viable yields on land unfit for food production. Several different types of climatic zones exist across India, so knowledge generated in one area is often not appropriate for other areas. Thus, knowledge transfer of *Jatropha* cultivation methods and their economics is yet another challenge (Wani and Chander 2012).
- There was no minimum support price or guaranteed purchasing for the *Jatropha* seeds. This was a problem since these kinds of supports were provided to many other commodities in India, putting biodiesel at a relative disadvantage. As a result, the price of *Jatropha* seeds was very high because most of them are used for plantation purposes rather than oil extraction. At this price, the manufacturing cost of biodiesel was three times the pump price of conventional diesel.
- Even though the consumption of edible oils in India was high, the availability of used cooking oil was very small, since it is typically reused until it disappears. Thus, there is no possibility to use waste cooking oil to produce biofuel in India.
- The use of lamp oil has been increasing rapidly in India, as there is no electrical power supply for 10–14 h a day in most rural areas. When the price of edible oils increases, people turn to the cheaper nonedible oils. The requirement of this

sector is more than 15 million tons (bio-kerosene). Since seeds can be collected and crushed, using hand-operated expellers, on a small scale in remote villages, the use of nonedible oils for lighting was rapidly expanding and creating a shortage of supply to the biodiesel industry.

- Most of the edible oils used are stable and do not decompose much in storage. Therefore these are preferred for the transesterification process. In contrast, nonedible oils are not very stable and require significant pretreatment with additional cost, so these are less preferred by the oil-producing companies.
- The cottage washing soap industry can use vegetable oils with a high content of free fatty acids. Since the prices of edible oils have doubled, many soap manufacturers in this unorganized sector are using nonedible oils since these are somewhat cheaper. This contributes to the supply shortage for biofuel producers.

8.3.3 Socioeconomic Impacts of Biofuels in India

When India's biofuel policy was adopted, one of the major motivations was to support social development through rural empowerment and development. The policy aimed to generate rural employment and achieve energy self-sufficiency and security in addition to environmental improvement. However, after a decade of efforts, Indian biofuel policies have contributed little toward these objectives.

Regarding rural employment generation, biodiesel was expected to contribute more than ethanol-based biofuels, using the National Rural Employment Guarantee Act (NREGA) program. Unfortunately, in most cases, the NREGA funds for biofuels were inadequate, either because money was allocated to competing government programs running in parallel in the same location, or because of bureaucratic problems in getting the funds to the right place at the right time. For the *Jatropha* plantation program, NREGA supported several initial activities for the first couple of years of the program but could not create enough interest among the farmers to continue in the program until seeds were available. As a result, a majority of the programs failed in the middle, and a large amount of money was wasted under this scheme.

It has become clear that *Jatropha* and other nonedible oil seed plantations need considerable regular agricultural care to cultivate it at an economic level of production, so the process is not cost-free. In addition, since *Jatropha* is a new crop, farmers also need new technical and economic knowledge to cultivate it effectively.

Over the last few years, all of the major *Jatropha* projects have produced significantly fewer seeds than planned, and quality has also been lower than expected. As a result, India's current oil extraction capacity of 600,000 t/day is running under 40% utilization, and plant operators are suffering large investment losses. The myth of *Jatropha* and other plant-based nonedible oil seeds as miracle crops for biodiesel has collided with reality in India. It is important to understand that these crops have to be recognized as regular and standard agricultural crops just like others. They incur production costs just like other crops, and they cannot be cultivated carelessly with little or no effort. Although the biofuel policy was supposed to support the rural landless laborers and farmers, in reality it supported partial employment for women and children mainly due to its irregularity and wage structure with below market rates. In most cases, it was observed that during the plantation process and seed collection, the involvement of rural women and children, for whom this was a part-time activity, reduced productivity. This further indicates the lack of incentives in the program to engage the rural male population.

Regarding the goal of promoting energy self-sufficiency and national energy security, it seems difficult for first-generation plant-based biofuels to make a significant contribution. With the burgeoning total fossil fuel demand in the country, the absolute amount of biofuels that would be required by the 20% target is also rapidly increasing and at a much faster rate compared to the pace of increase in biofuel production. Finally, the goal of improving environmental quality through biofuels remains a lower priority, as large-scale market production has not been achieved.

Uneven availability of market information, which is related to underdeveloped regulation, is another problem. The majority of the market information still lies with the downstream stakeholders starting from the seed crushers to the oil marketing companies. However, a severe lack of information still persists among the upstream stakeholders including the farmers and field workers. Such information asymmetries have created opportunities for middle traders in the market who are distorting the pricing system. It has been recorded that in some places *Jatropha* seeds are sold ten times the market price to the mill owners, while the farmers and producers are still getting a below average price (even lower than the minimum selling price).

India's national biofuel policy and its mission were well-intentioned, but many details were not developed, so they were not well-implemented. Many aspects of the policy were either vague or not well developed, especially in comparison with other industrial promotion policies, particularly related to pricing.

8.3.4 Analysis

As the first-generation biofuels have come under global scrutiny in the context of their sustainability in terms of net energy gain, emission reduction potential, and resource utilization, the Indian biofuel program has also not been free from those concerns. India has been suffering from a severe water crisis and lack of irrigation facilities. India's bioethanol production is highly vulnerable to water shortages since it is heavily dependent on water-intensive sugarcane production. Sugarcane is one of the most water- and energy-intensive crops, and unfortunately in India, sugarcane is being produced in the most water-stressed regions and with complete groundwater irrigation. Given the limited availability of natural resources in India, especially land and water, it is doubtful that the country can produce enough surplus sugarcane in the coming years to satisfy the potentially huge demand for ethanol.

Moreover, the land categorized as wastelands designated for nonedible oil seeds production is either available only in remote locations or above 1500 m in altitude. Wastelands in either of these cases would be unsuitable for oil seed production and its commercial utilization. Remoteness of location would create huge additional expenses for transportation of saplings, seeds, and human resources as well as hamper the regular maintenance of the trees which is essential to achieve a minimum acceptable seed yield.

Finally, for a country like India, first-generation biofuels are still a luxury in the sense that India still has severe food shortages and millions of people are suffering from malnutrition. Every effort should be made in India to produce more foods and edible oils by utilizing every piece of land. However, alternative sources of bioenergy could be explored such as algae. In India, algae-based biofuel production research has been conducted for a long time, but it needs continuous encouragement from the government as well as from the industries to make it faster. It is also not clear how much land and water will be required.

8.4 China

8.4.1 Overview of China's Main Policies: Promotion of Renewable Energy

In 2010, China was the second largest energy-consuming country in the world (EIA 2010). The majority of China's primary energy came from abundant domestic coal to meet domestic demand, not only for households but also industrial use (Martinot and Junfeng 2007; Zhang and Siang 2007). In response to its rapid increase in energy use, the nation has made a major effort to gear up its use of renewable energy. As of 2007, China received only 8% of its primary energy from renewable energy, and its target shares were set at 10% and 15% by 2010 and 2020, respectively (NDRC 2007). To meet these ambitious goals, China enacted the Renewable Energy Law in 2005. This law has several objectives including improving energy structure, diversifying energy supplies, safeguarding energy security, protecting the environment, and realizing the sustainable development of economy and society, and it covers a comprehensive list of renewable energy sources. Short-term (2010) and long-term (2020) renewable energy targets are summarized in Table 8.4. China's renewable energy policies stress the large-scale provision of electricity nationwide in the midst of rapid industrialization. As of 2010, biofuels' contribution as a renewable energy source was relatively small in China—the large majority of investment in renewable energy was for wind power (70%), followed by other renewables (17%) and solar (8%), and biofuels accounted for only 3.6% (Pew Charitable Trusts 2010).

Table 8.4 China's targets for annual renewable energy utilization and supply in 2010 and 2020

Form Source	Source	Electricity utilization (share in total electricity utilization, %)	Gas/heat supply	Biofuel utilization
Short term (2010)[a]	Non-biomass source	Hydro: 190 GW (92.3%)	Heat supply by solar and geothermal: 100 million J	
		Wind: 10 GW[c] (4.9%)		
		Solar: 0.3 GW (0.1%)		
	Biomass source	Biomass (solid and gas): 5.5 GW (2.7%)	Biogas supply: 19 billion cubic meters	Bioethanol from nonedible food sources: 2 million tons
			Solid biomass: 1 million ton	Biodiesel: 0.2 million tons
Medium and long term (2020)[b]	Non-biomass source	Hydro: 300 GW (82.9%)	No information	
		Wind: 30 GW (8.3%)		
		Solar: 1.8 GW (0.5%)		
	Biomass source	Biomass (solid and gas): 30 GW (8.3%)	No information	Bioethanol from nonedible food sources: 10 million tons
				Biodiesel: 2 million tons

Source:
[a]Developed based on the Renewable Energy Development Plan for the 11th Five-year Period (NDRC 2008)
[b]Developed based on the Medium- and Long-Term Development Plan for Renewable Energy in China (NDRC 2007)
Notes:
[c]5 GW in the Medium- and Long-Term Development Plan for Renewable Energy in China

8.4.2 *Overview of Main Biofuel Market Conditions in China*

8.4.2.1 Bioethanol

China's bioethanol production in 2007 was the third largest in the world at 1.33 million tons (Huang et al. 2008). Estimated 2008 production totaled 1.55 million tons, of which 1.42 million tons of bioethanol were derived from corn and wheat produced at four designated plants operating at almost full capacity (84–100%) (USDA 2008). The remaining 130,000 tons of bioethanol came from cassava whose plant operates at only 65% of its capacity (USDA 2008).

Bioethanol production initially utilized old grains in stock. The production was mainly from corn and partially from wheat at four designated state-owned plants in Heilongjiang, Jilin, Henan, and Anhui provinces. However, in May 2007, bioethanol production from corn and wheat was capped by the government, which stopped approving new bioethanol production from food for fear that food-based ethanol production would cause food prices to increase (Sun 2007; Huang et al. 2008).

To supplement biofuel production, cassava was identified as one of the most promising nonfood feedstocks to produce bioethanol. In 2008, the government approved a new state-owned facility to produce bioethanol from cassava in Guangxi province (USDA 2008). The province was once known for its large-scale cassava-producing region, but the cassava industry was suffering from low prices for fresh cassava as well as starch. Therefore, biofuel production from cassava was expected to create employment and improve livelihoods in the region. However, the production of cassava has not been enough to meet domestic demand, and actually the bioethanol company imported feedstock from Thailand and Vietnam (GSI 2008; USAID 2009). The reality is that the majority of bioethanol will be produced from corn and wheat for the near future (GSI 2008).

Bioethanol blending mandates have been implemented in ten provinces, including one autonomous region. Province-wide blending mandates (E10) were first introduced in 2005 in five provinces (Heilongjiang, Jilin, Liaoning, Henan, and Anhui) which have bioethanol plants located in or near the province (GTZ 2006). Blending mandates then expanded to additional cities in four neighboring provinces (Hebei, Hubei, Shandong, and Jiangsu). In April 2008, after the government capped bioethanol production from food, Guangxi Zhuang Autonomous Region became the tenth province to introduce province-wide blending mandates. It was the first case of ethanol production from cassava (People's Daily Online 2008).

8.4.2.2 Biodiesel

The total volume of biodiesel produced in 2007 was reported at 300,000 tons, which was on a smaller scale compared to bioethanol (ERI 2008; USDA 2008; F.O. Licht 2009). There were a dozen operating plants using waste oil as a feedstock and 20 planned plants which will operate on not only waste oil but also other feedstocks such as *Jatropha* as of 2008 (Huang et al. 2008; Morimoto 2008; USDA 2008). The production capacity of each plant is relatively small due to an insufficient supply of feedstock. China is a net importer of vegetable oil, and there are difficulties in feedstock collection and marketing (Huang et al. 2008; USDA 2008).Unlike bioethanol, there are no blending mandates for biodiesel. There are voluntary standards for 100% biodiesel (JIE 2008; USDA 2008), and a standard for 5% (B5) was introduced in 2010.

Nevertheless, a number of large companies planned to invest in biodiesel production. In 2008, the NDRC approved three state-owned plants to produce biodiesel from *Jatropha* to be implemented by either PetroChina or Sinopec in Sichuan, Guizhou, and Hainan provinces. Out of 32 plants (both operating and planned),

Table 8.5 Changes in agricultural labor availability in China

Labor	2000	2005
Rural labor [1000 persons]	479,821	504,050
Indexed rural labor change (2000 = 100)	100	105.0
Agricultural labor [1000 persons]	327,975	299,755
Indexed agricultural labor change (2000 = 100)	100	91.4
Indexed agricultural labor share change (2000 = 100)	100	87.0

Source: China Statistical Yearbook 2007

seven plants were operated by China's largest biodiesel producer Gushan Environmental Energy Limited based in Hong Kong (Morimoto 2008; PetroChina 2008). Some biodiesel feedstocks, industrial waste oil and palm oil, have been imported from Malaysia (PetroChina 2008).

Jatropha was regarded as one of the most promising feedstocks for biodiesel. In 2007, the State Forestry Administration (SFA) and PetroChina signed a contract to cooperate on a 40,000 ha *Jatropha* project in Yunnan and Sichuan and with COFCO (China National Cereals, Oils and Foodstuffs Corporation) in Guizhou. Foreign investment also flowed in from the UK to Guangxi and Yunnan, from the USA to Sichuan, and from Germany to Yunnan (Mang 2008).

8.4.2.3 Emerging Research on Second-Generation Biofuels

With abundant agriculture and forestry residues available in China, a considerable amount of second-generation biofuels was expected. However, there were only two second-generation biofuel pilot plants operating using corn stover as feedstock as of 2010 (IEA 2010). Water use and wastewater for/from the process could potentially cause environmental problems (IEA 2010).

8.4.3 Socioeconomic Impacts

8.4.3.1 Employment

Agricultural labor availability in rural China has been rapidly decreasing, and more labor has been absorbed by non-agricultural sectors as the nation's economy developed (see Table 8.5). According to one estimate, biofuels were predicted to create more than nine million jobs in China (Dufey 2006). The NDRC estimated that 1,000 people could be hired at a 100,000 ton-scale ethanol plant (GSI 2008).

In the case of *Jatropha* production, potential labor shortages could become more severe if more labor is needed to harvest in the future when *Jatropha* trees mature. The additional labor needed for harvesting *Jatropha* might be diverted from food

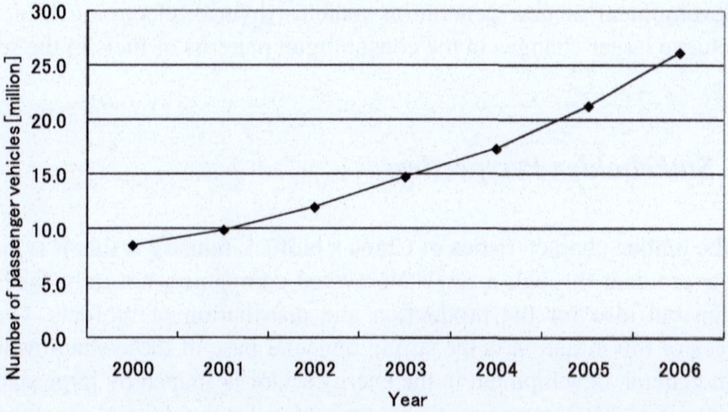

Fig. 8.4 Number of passenger vehicles in China (Source: National Bureau of Statistics and NDRC 2007)

crops, and this in turn could lead to a shortage in food production—a possible two-step food–fuel conflict (Sano et al. 2012).

8.4.3.2 Rural Development

To what extent biofuel production could contribute to rural development depends on whether or not the rural economy can supply sufficient feedstock to biofuel producing factories. In this sense availability of inputs for production such as natural resources and labor mentioned above are crucial. Water is one of the crucial inputs, and potential shortages are an important concern.

In order to generate additional income for rural households, business coordination between a large number of farmers and a few state-owned enterprises and biofuel processing firms would be important. For overall improvement of welfare in rural communities, however, liquid biofuels in general may have a smaller direct contribution compared to potential alternatives, because they have fewer other applications besides use in transport sector unlike other forms of biomass utilization such as biogas and solid biomass. For instance, solar and biogas cookers could lower the energy expense for households according a case study conducted in rural region of Gansu province (Li et al. 2009).

8.4.3.3 Energy Security

Biofuels might be able to make some contribution to the diversification of energy forms in the transport sector; however, the extent is expected to be limited since the rapid increase in vehicle ownership (see Fig. 8.4) is likely to be higher than the potential for expansion of biofuels. Thus, by themselves, biofuels would have negligible effect in reducing China's oil consumption or energy security (GSI 2008).

Rapid development of new-generation vehicles (hybrid, electronic vehicles, etc.) might achieve larger changes in the consumption patterns of fuels in the sector.

8.4.4 Stakeholder Perspectives

One of the unique characteristics of China's biofuel industry is that it is dominated by the government through a few state-owned companies, not only for feedstock production but also for the production and distribution of biofuels. One of the advantages of this situation is the strong financial base of these state-owned enterprises. In general, development in the energy sector is shaped by large state-owned companies which have much greater investment and technological capabilities compared to small- and medium-sized companies, and this is also case in the bioenergy (Gan and Yu 2008). In 2006, PetroChina provided five million RMB to initiate four demonstration projects in Yunnan (ICRAF China 2007). Another advantage is that state-owned companies can manage supply chains more easily. In this sense, standard setting and implementation would be also relatively easier.

On the other hand, the biggest disadvantage is that the market is relatively closed and dominated by a few companies, making the market more uncompetitive and inefficient. Bioethanol for fuel is not a market-driven segment of the economy, and there are only a few licensed companies. In addition, the pricing regime discourages the private sector's investment in fuel ethanol production and ensures limited competition for existing producers (GSI 2008; Huang et al. 2008; USDA 2008). This situation may cause technological innovation by the private sector to be slow. Also, there is a high probability that related decision-making by the central government does not fully consider local conditions or implications for local economies. Energy policies are under the jurisdiction of the Energy Bureau of the NDRC, which has a higher position than other bureaus in NDRC's internal hierarchy, but it is heavily influenced by large energy-related state-owned companies (Takamizawa 2009).

8.4.5 Analysis

Although China's biofuel production is relatively large on a global scale, it has a relatively smaller role in renewable energy promotion within China itself. Moreover, in China, biofuel promotion tends to be more closely related to agricultural policies than renewable energy or climate change policies. China, as one of the largest grain producers in the world, made a timely policy response to address food–fuel conflict concerns in 2007. Partially because of the government's strong grip on both biofuel production and distribution, a significant food–fuel conflict feared by many researchers has been avoided. However, this has dampened the high hopes for biofuel promotion in China.

Still, China has made advances in feedstock diversification for the first-generation biofuels (*Jatropha*, cassava, sweet sorghum, etc.), invested in the development of second-generation biofuels, and explored production outside its territory, for example, the potential for palm oil plantations in Africa. In order to meet its skyrocketing energy demand, China must continue to explore all forms renewable energy, even those with a relatively smaller scale. China has a large potential for second-generation biofuels (Eisentraut 2010). Second-generation biofuels could play a more significant role as related technologies become more advanced, more capital becomes available, especially including overseas investment, and associated potential problems such as water scarcity are solved.

Biofuel production calls for close attention to the local conditions because natural resource availability, especially water and land, suitable agricultural/farming technologies, and socioeconomic conditions vary greatly across locations. Knowledge and assessment of local biofuel producing conditions are essential. Attention to labor availability is also important considering the increasing numbers of migrant workers and aging workers in the rural labor market.

More opportunities may arise for biofuels to contribute to sustainable development if the scope of biofuel industries expands to explore by-products, diversified products, or alternative feedstocks, including second generation. These would create more options for local economies. For instance, residues from *Jatropha* production can be used as fertilizers or for pest management, and glycerol produced during transesterification as a by-product can be used for soaps and lubricants (ICRAF China 2007). The use of biodiesel for rural electrification may not be as relevant to China compared to other developing countries, since the country has already achieved over 98% electrification in rural areas (Jiahua et al. 2006). Still, biodiesel could be used for grinding wheat (ICRAF China 2007) or as an alternative to coal or firewood, helping to reduce indoor air pollution, labor needed to collect fire wood, and the threat of deforestation. The cassava industry could start selling diversified starch-based products in the market as well, although this could potentially affect feed markets. The government's role in supporting R&D would be critical if China continues to rely on state-owned enterprises. Thus, state-owned enterprises have a critical role to play in influencing the socioeconomic impacts of biofuels.

8.5 Japan

8.5.1 Overview of Japan's Main Policies

Japan started to promote biofuels from the mid-2000s by setting national strategies and plans to promote biofuels including the "Biomass Nippon Strategy"[1] (2002, revised in 2006), the "Kyoto Protocol Target Achievement Plan"[2] (2005), and the "New National Energy Strategy"[3] (2006).

[1] "Baiomasu Nippon Sogo Senryaku".

[2] "Kyoto Giteisho Mokuhyo Tassei Keikaku".

[3] "Shin Kokka Enerugi Senryaku".

Table 8.6 Targets for bioethanol to be introduced by oil refiners (to meet the requirements of the Law for the Sophisticated Structure of Energy Supply)

FY 2011	FY 2012	FY 2013	FY 2014	FY 2015	FY 2016	FY 2017
210,000 kL	210,000 kL	260,000 kL	320,000 kL	380,000 kL	440,000 kL	500,000 kL

(In crude oil equivalent)

The specific short-term numerical target related to biofuel introduction was set at 500,000 kL in oil equivalent by 2010, incorporated in both the Kyoto Protocol Target Achievement Plan and the revised Biomass Nippon Strategy. For the period after 2010, the targets for biofuel introduction were set in the Basic Energy Plan in 2010. Its midterm target by 2020 intended to increase the share of bioethanol in gasoline to more than 3% nationwide, with the conditions that GHG emissions should be reduced sufficiently and economic viability should be ensured. The Plan further aimed to increase the use of biofuels to the maximum extent by 2030 using next-generation biofuel technologies such as biofuels from cellulosic materials and algae.

A roadmap was published in 2010 which requested oil refiners to introduce 500,000 kL of ethanol (in crude oil equivalent) by 2017 (Table 8.6). This roadmap takes the 2020 target into consideration and aims to implement the "Law to Promote Utilisation of Non-fossil Fuel Energy Sources and Efficient Use of Fossil Energy Raw Materials by Energy Suppliers" ("Law for the Sophisticated Structure of Energy Supply")[4], which was enacted in 2009 and required energy suppliers to promote biofuels and biogas as non-fossil energy, assuming that biofuels can reduce GHG emissions by more than 50%.

Other policies to promote biofuels include an import tax exemption on ethyl tertiary-butyl ether (ETBE), a fuel tax exemption for bioethanol, and various financial and tax support measures for the producers of feedstocks and biofuels.

8.5.2 Overview of Main Biofuel Market Conditions

In the area of domestic production, the roadmap of *the Large-Scale Expansion of Domestic Biofuel Production* set a target to produce 50,000 kL of ethanol (30,000 kL in oil equivalent) domestically by FY 2011. Financial support for pilot projects and research and development (R&D) of advanced biofuels also has been provided by relevant ministries including the Ministry of the Environment (MOE), the Ministry of Agriculture, Forestry and Fisheries (MAFF), and the Ministry of Economy, Trade, and Industry (METI). The nationwide annual production was approximately 15,000 kL as of the end of the fiscal year (FY) 2009, increasing from 200 kL in the

[4] "Enerugi Kyokyu Jigyosha ni yoru Hi-kaseki Enerugi-gen no Riyo oyobi Kaseki Enerugi-genryo no Yuko na Riyo no Sokushin ni kansuru Horitsu"("Enerugi Kyokyu Kozo Kodo-ka Ho").

previous FY (MAFF 2009).[5,6,7] Feedstocks used in domestic production vary from edible crops (high-yielding rice, substandard flour, sugar beets, etc.) to waste materials (construction waste timber, saw mill waste, food waste, etc.). Major fuel ethanol pilot projects in Japan as of FY 2008 are listed in Table 8.7.

In contrast, biodiesel production in Japan has not been mainstreamed into the national policy. Production of biodiesel has been mainly based on waste cooking oil, through projects carried out by local governments or nongovernmental organizations. The total amount of biodiesel production as of March 2008 was estimated at 10,000 kL, which was double the amount from the previous year (MAFF 2009). A few examples of biodiesel utilization on a relatively larger scale are found in Kyoto City, Toyama City, and Iwaki City (Fukushima Prefecture) and Shiogama City (Miyagi Prefecture).

Sales of bioethanol-blended gasoline were started in 2007. The number of service stations retailing ETBE-blended gasoline was 1,710 as of 10 December 2010. In contrast, the number of service stations selling E3 was 18 in Osaka Prefecture and 6 in Kanagawa, Chiba, Ibaraki, and Aichi. Even if the Kyoto Target Achievement Plan could be achieved, this would amount to approximately 1% of gasoline consumption.[8]

8.5.3 Socioeconomic Impacts

The Biomass Nippon Strategy envisions the socioeconomic benefits of biomass utilization would be in the areas of contribution to the creation of a sound material-cycle society, incubation of new industries, revitalization of rural economies, and global warming mitigation. This section discusses the impacts of biofuels relating to a sound material-cycle society, rural development, and energy security.

The promotion of biofuels derived from unutilized materials and wastes is expected to enhance material recycling in resource-poor Japan. In fact, waste utilization has played an important role in biodiesel production in Japan through projects carried out by local governments or nongovernmental organizations to collect waste cooking oil and mix it with diesel fuel. Projects to produce waste-based ethanol also have been launched in some areas of Japan, utilizing materials such as food waste and waste construction timber. Data shows that there is still a significant amount of unutilized biomass which could be converted to ethanol. However, there are challenges related to difficulties in collection from small-scale waste generators. In addition, especially in the case of construction waste timber, competition with

[5] The exact amount of production by each company is not published.

[6] The Japanese fiscal year starts on 1 April and ends on 31 March.

[7] Koji Okura, Deputy Director of the Biomass Policy Division, MAFF, replied to the question by the author at the Biomass Expo 2010, 18 November 2010.

[8] 500,000 kL in oil equivalent is 561,797.8 kL in gasoline, and the actual gasoline consumption is 2008 was 57,473,000 kL.

Table 8.7 Major pilot projects for fuel ethanol in Japan

Area	Implementer	Related ministry	Project outline
Shimizu Town, Hokkaido	Hokkaido Bioethanol Co. Ltd.	MAFF	Production from sugar beets, flour, etc.
Tokachi Area, Hokkaido	Tokachi Area Promotion Organisation	MOE, MAFF, METI	Production from substandard flour, corn, etc. and demonstration of gasoline blended with 3% ethanol (E3)
Tomakomai, Hokkaido	Oenon Holdings, Inc.	MAFF	Production from rice, etc.
Shinjo City, Yamagata Prefecture	Shinjo City	MAFF	Production from sorghum and E3 demonstration
Niigata City, Niigata Prefecture	National Federation of Agricultural Cooperative Associations	MAFF	Production from rice and E3 demonstration
Kanto Region	Petroleum Association of Japan (PAJ)	METI	Demonstration of ETBE
Sakai City, Osaka Prefecture	Bioethanol Japan Kansai, Osaka Prefecture	MOE	Production from construction waste timber and E3 demonstration
Maniwa City, Okayama Prefecture	Mitsui Engineering & Shipbuilding Co. Ltd, Okayama Prefecture, Maniwa City	METI	Production from lumber waste, etc., and E3 demonstration
Kitakyushu City, Fukuoka Prefecture	Nippon Steel Engineering Co. Ltd.	METI, MOE	Production from food waste and E3 demonstration
Ie Island, Okinawa Prefecture	Asahi Breweries, Ltd., National Agricultural Research Center for Kyushu Okinawa Region (KONARC)	MOE, MAFF, METI, Cabinet Office	Production from molasses with a high biomass amount and E3 demonstration
			Production from molasses and E3 demonstration
Miyakojima Island, Okinawa Prefecture	Ryuseki Corporation	METI, MOE, MAFF, Ministry of Land, Infrastructure, Transport and Tourism (MLIT), Fire and Disaster Management Agency, Cabinet Office	

Source: Committee for Eco-fuel Utilisation Promotion (2008a, b)

other uses has intensified as the wood chip market has experienced drastic fluctuations due to an increase in demand for biomass energy and a reduction in supply of construction waste timber due to stagnation in the construction market (Matsumoto and Sano 2011).

The effects of biofuel crop production on rural development would depend on which crops are cultivated in the future and the location where they are planted. In 2005, the area of "abandoned cultivated lands" (lands which are no longer being cultivated) was 386,000 ha, which is equivalent to 9.7% of total cultivated land (the sum of cultivated lands under management and abandoned cultivated lands) (Saigo 2008). Utilization of such abandoned cultivated lands as well as marginal lands could bring opportunities for rural development.[9]

The potential of biofuels to improve energy security seems very limited. In 2011, Japan's production target was much smaller than its introduction target. For example, for FY 2011, the government aimed to increase biofuel production up to 50,000 kL (30,000 kL of oil equivalent from both bioethanol and biodiesel).[10] According to the roadmap to achieve the Basic Energy Plan, the targeted amount of bioethanol introduction for that year was 210,000 kL in crude oil equivalent. This indicates that even if the Japanese producers could successfully achieve the targeted level of production, it is far short of the targeted level of introduction, and the rest would need to be imported. It could be argued that biofuel imports might contribute to energy security by diversifying the energy sources and supplying countries, considering the fact that about half of Japan's total energy supply comes from imported oil, of which almost 90% is imported from the Middle East, and that the transport sector is almost entirely dependent on oil. However, potential suppliers of bioethanol are limited to a few countries, and Brazil is currently regarded as the only country with the potential capability to export significant quantities in a stable manner. In addition, when the GHG reduction potential is considered, Brazil is the only foreign supplier which could have some possibility to reduce GHG emissions by more than 50%.

8.5.4 Stakeholder Perspectives

8.5.4.1 Government

As biofuels encompass several different policy areas, including agriculture, energy, industry, and environment, various government ministries have introduced related national strategies, plans, and policies. For example, the Biomass Nippon Strategy

[9] For example, there is a rural revitalization project in Ibaraki Prefecture involving the cultivation of sweet sorghum in abandoned agricultural land to produce bioethanol.

[10] Specified in the roadmap entitled the "Large-Scale Expansion of Domestic Biofuel Production" (Kokusan Baionenryo no Ohaba na Seisan Kakudai) (Biomass Nippon Strategy Promotion Committee 2007).

is an initiative of the Ministry of Agriculture, Forestry and Fisheries (MAFF) in cooperation with other ministries. MAFF promoted increased domestic production of bioethanol, with a strong emphasis on the technology development in the area of soft cellulose. The Ministry of the Environment (MOE) established a Committee for Eco-fuel Utilisation Promotion and promoted pilot projects to introduce E3. The Basic Energy Plan, which set a target to increase the share of bioethanol in gasoline to more than 3% by 2020, was developed by the Ministry of Economy, Trade, and Industry (METI). As shown in Table 8.2, those ministries have supported production projects for fuel ethanol from various feedstocks, independently in some cases and jointly in others.

In the area of introduction, that is, blending ethanol into transport fuel (especially gasoline), promotion policies were introduced without a full agreement between the MOE and the Petroleum Association of Japan (PAJ) on the blending method: whether ethanol should be directly blended or should be first processed into ethyl tertiary-butyl ether (ETBE) and then blended. This led to two different markets of ethanol-blended gasoline, one for E3 and one for ETBE-blended gasoline (so-called biogasoline). The lack of the agreed national blending policy was noted in the screening process to reduce the national budget in 2010 and as a result the MOE's budget related to E3 promotion was recommended to be halved.

8.5.4.2 Oil Industry

PAJ was requested by the government to increase the introduction of biofuels to 210,000 kL in oil equivalent (840,000 kL in bio ETBE) as a part of the effort to achieve the Kyoto Protocol Target Achievement Plan (a total of 500,000 kL in oil equivalent for liquid transport fuel), and it is likely to achieve the goal. However, the oil industry has opposed large-scale introduction of biofuels for several reasons such as limited supply, concerns about the stability of supply, expected high infrastructure investment costs (such as oil refineries), and the potential for food–fuel conflict. PAJ insisted on waiting for the commercialization of production technology before discussing the expansion of biofuel introduction.[11]

8.5.4.3 Automobile Industry

Many Japanese auto manufacturers have already started exporting E10-compliant vehicles, and manufacturers such as Toyota, Honda, and Nissan have already been selling new vehicles compatible with E10 (Sakata 2009). In addition, some companies have already launched sales of flex-fuel vehicles in Brazil, and vehicles compatible with E85 in the United States, and E20 in Thailand. The Japan Automobile Manufacturers Association (JAMA) published its position statement on both ethanol-blended gasoline and FAME-blended diesel and stated that it has

[11] Presentation made by the PAJ on the Medium- and Long-Term Roadmap on 3 June 2010.

consistently supported the use of biofuels complying with appropriate sustainability criteria as part of an integrated approach to the reduction of CO2 emissions. However, it also emphasized the need to ensure that biofuels are equivalent in quality to conventional fuels so as to achieve satisfactory safety and emission performance of vehicles. It also emphasized the need for clear and harmonized fuel quality standards.[12,13]

8.5.4.4 Consumers

The results from an annual website questionnaire survey conducted by the PAJ in 2010 indicated that the image of biofuels had turned more positive, compared to the one conducted in 2008 when global food prices soared. In 2010, 63% of 4390 respondents supported the statement that "use of biofuels for transportation should be promoted if it is within the range that does not affect other issues such as the food problem," which was a 6.5% decrease from the previous year. In comparison, the ratio of respondents who replied that "I support the proactive promotion of biofuels in order to prevent global warming" increased by 4% from the previous year and reached 29.9%.

8.5.5 Analysis

Although the domestic production of biofuels has been increasing, the ability of biofuels to contribute to Japan's energy security is constrained by the potential scale of domestic production and availability of imports. In contrast, biofuels might play a more significant role in the revitalization of rural economies and the development of a sound material-cycle society (Matsumoto et al. 2009). The success of such efforts relies on the future development of technologies and socioeconomic infrastructure.

The introduction targets that the oil refiners have been requested to meet (from FY 2011 to 2017: see Table 8.1) are larger than the scale of domestic production. Thus, Japan will need to continue to import a significant amount of biofuels at least for the next decade. Under these circumstances, it is necessary to set appropriate sustainability criteria for biofuels. The Japanese government has been in the process of developing such a standard and examined a 50% GHG reduction as a criterion.

[12] JAMA Position Statement, FQ-01, 2009.10.30 "Quality of Bioethanol and Use of Ethanol-blended gasoline".

[13] JAMA Position Statement, FQ-02, 2009.10.30 "Quality of Biodiesel (FAME) and Use of FAME-blended diesel".

8.5.6 Policy Implications

Considering the limitation of feedstock production and the state of ethanol production technologies, it seems reasonable to maintain the modest introduction target. In the area of revitalization of local economies through the promotion of biofuels, decisions on the location of cultivation and the choices of energy crop species are crucial. For a sound material-cycle society to be realized, although second-generation biofuel production technologies to utilize rice straw and unutilized woody biomass are being advanced, further development is necessary to reduce production costs and make them commercially viable. In addition, developing efficient collection systems including small-scale waste generators is crucial. Finally, setting appropriate sustainability criteria would be especially important in Japan as it needs to import ethanol from overseas to meet the introduction targets.

8.6 Conclusion

Major biofuel promotion policies in the case study countries started from the mid-2000s and had largely similar objectives, although with different emphases. All four countries emphasized rural development, but Japan placed comparatively more emphasis on the goal of reducing GHG emissions, while the other three countries placed more emphasis on energy security. Somewhat surprisingly, several major aspects of biofuel policies converged among the four countries, despite significant differences in their situations. The initial biofuel targets set by Indonesia and India were overambitious, but these countries have since backed off of these targets, while those of Japan and China were more conservative from the early stages. Partly, this reflected the now widespread sensitivity among governments about the potential for biofuels to cause a food–fuel conflict. The governments of all four countries have been very sensitive to this issue. Biofuel promotion policies in Indonesia and India in particular tended to focus on promoting specific biofuel feedstocks, but later all four countries recognized that overdependence on one or a few feedstocks is not desirable. In all cases, the biofuel boom of the 2000s was supported by high oil prices, and the subsequent oil price fall and global financial crisis severely harmed the economic viability of biofuels. Nevertheless, governments of all four countries, albeit to different extents, have engaged in research and testing of alternative feedstocks and second-generation biofuels. Finally, all four countries have recognized the limitations of biofuels for energy security and placed more emphasis on their potential to contribute to rural development.

Biofuels may have some potential to contribute to rural development, even in developed countries such as Japan. However, the case studies in China, India, and Indonesia showed that biofuels are not likely to be a "miracle solution" to promote rapid rural development, and the idea of growing nonfood crops on wastelands is too good to be true. Much "wasteland" would need significant inputs of fertilizer and water in order to produce a significant quantity of biofuels. In any case it is not clear how much wasteland actually exists, and often it is actually being used for some other economically valuable purpose, especially by lower income people, or providing ecosystem services. Farmers have various crop alternatives, and biofuel crops, especially nonfood crops with limited alternative uses, are often not very attractive options without significant economic support, which governments have been reluctant to provide.

Regarding energy security, these case studies of relatively large countries show that the ability of biofuels to contribute to energy security could be modest but is fundamentally limited. Biofuels may contribute to supply diversification to some extent. However, even the achievement of modest targets in China, Japan, and India will require imports. In Indonesia, despite the ambitions of some for the country to become the "Middle East of biofuels," the main large-scale crop, palm oil, is too important for food purposes for the government to allow its significant diversion to other uses, and this was the case even before the biofuel boom. It is already a significant challenge for Indonesia to produce enough fuel domestically to meet its targets. Moreover, other than Brazil, potential sources for imports are unclear.

Biofuels do seem useful for recycling waste materials, especially in Japan, although in some cases, biofuels compete with alternative uses for recycling the wastes. In developing countries like India, cooking oil is often reused until it disappears, so other waste sources would have to be considered.

Several policy implications can be drawn from these cases. First, it may be desirable to adopt a cautious stance and avoid setting high unrealistic targets. Large-scale, rapid expansion of biofuels could pose high risks of food–fuel conflict and may not be feasible due to limited supplies of land, water, and labor. If targets cannot be met by domestic production, imports would be necessary. Too high targets risk encouraging unsustainable production, deforestation, water shortages, food–fuel conflict, and inappropriate appropriation of land used by poor people. Modest targets, near existing utilization rates, may be more sustainable.

Second, all countries dealt with the question of how much biofuels should be promoted through special economic incentives such as subsidies, mandatory targets, or price regulations. This is an especially important issue in countries like India and Indonesia, where many sectors receive special treatment—particularly fossil fuels, which is the main sector competing with biofuels. Therefore, a lack of special promotion measures becomes in effect a disincentive policy, so the governments of India and Indonesia in particular have been under strong pressure from businesses interested in promoting biofuels to adopt these kinds of measures. In principle, such measures could be justified if biofuels provide important social benefits, but since these benefits have been shown to be still unclear, the caution displayed by India and Indonesia seems justified. To be sure, subsidies and other special promotion measures for fossil fuels are also problematic from the point of view of environment and sustainability (UNEP 2008), and reducing them is widely viewed as beneficial, but nevertheless, reducing fossil fuel subsidies would serve to make agricultural prices more closely linked to fuel prices in these countries and create a more level playing field with biofuels.

Finally, it is desirable to promote the use of sustainability standards, given the remaining large uncertainties about the impacts of biofuel production and availability of inputs such as land and water, the great variation in local conditions, and the likelihood that biofuels will be globally traded. Standards can enable individual biofuel stakeholders to demonstrate that their particular production methods in their particular circumstances is sustainable. To be sure, these standards have various limitations, but sustainability standards seem to be the main possibility to demonstrate the potential for biofuel sustainability on a case-by-case basis, taking into account local conditions.

References

APEC (2008) The future of liquid biofuels for APEC economies. Ed. Energy Working Group, Asia-Pacific Economic Cooperation (APEC) Secretariat, Singapore. https://www.nrel.gov/docs/fy08osti/43709.pdf

Biomass Nippon Strategy Promotion Committee (2007) Kokusan baio nenryo no ohaba na seisan kakudai [Large scale expansion of domestic bio-fuel production]

China ICRAF (2007) Bilfuels in China: an analysis of the opportunities and challenges of *Jatropha curcas* in Southwest China. World Agroforestry Centre China, Beijing

Committee for Eco-fuel Utilisation Promotion (2008a) Status of biofuel projects in Japan (in Japanese), Handout 2–1 of the Committee meeting

Committee for Eco-fuel Utilisation Promotion (2008b) Status of measures related to eco-fuels (in Japanese), Handout 2–2 of the Committee meeting

Commodity Online (2008) Indonesian biofuel export hurt by legislation. Commodity Online, 20 March

Dufey A (2006) Biofuels production, trade and sustainable development: emerging issues. IIED (International Institute for Environment and Development), London

EIA (2010) China energy profile. http://www.eia.doe.gov/country/country_energy_data.cfm?fips=CH#. Accessed 27 Dec 2010

Eisentraut A (2010) Sustainable production of second-generation biofuels: potential and perspectives in major economies and developing countries. International Energy Agency, Paris, France, Available at: https://www.iea.org/publications/freepublications/publication/second_generation_biofuels.pdf

ERI (2008) Transition for non-food feedstock in the Chinese biofuel industry: current status and challenge. Policy dialogue on biofuels in Asia: benefits and challenges, Beijing, China, United Nations Economic and Social Commission for Asia and the Pacific

Fargione J, Hill J, Tilman D, Polasky S, Hawthorne P (2008) Land clearing and the biofuel carbon debt. Science 319(5867):1235–1238

Gan L, Yu J (2008) Bioenergy transition in rural China: policy options and co-benefits. Energy Policy 36(2):531–540

GSI (2008a) Biofuels – at what cost? Government support for ethanol and biodiesel in China. Global Subsidies Initiatives of the International Institute for Sustainable Development, Geneva

GSI (2008b) Biofuels, at what cost? (Government support for ethanol and biodiesel in Indonesia). Global Subsidies Initiative of the International Institute for Sustainable Development, Geneva

GTZ (2006) Liquid biofuel for transportation: Chinese potential and implications for sustainable agriculture and energy in the 21st century. GTZ, Beijing. http://www.commodityonline.com/commodities/oil-oilseeds/newsdetails.php?id=6583. Accessed 6 May 2008

Huang J, Qiu H et al (2008) Strategies and options for integrating biofuel and rural renewable energy production into rural agricultural for poverty reduction in the greater Mekong subregion: a case study of China. ADB, Manila

IEA (2008) Energy policy review of Indonesia. OECD/IEA, Paris

IEA (2010) Sustainable production of second-generation biofuels potential and perspectives in major economies and developing countries. OECD/International Energy Agency (IEA), Paris

Jayawardena M (2005) Electricity for all: options for increasing access in Indonesia. Asia Sustainable and Alternative Energy Program (ASTAE). World Bank, Washington, DC. http://documents.worldbank.org/curated/en/725981468025507664/Electricity-for-all-options-for-increasingaccess-in-Indonesia

JIE (2008) Asia biomass handbook (Japanese). Japan Institute of Energy, Tokyo

Kusdiana D (2006) Recent updates in biofuel development in Indonesia. Presentation by Ministry of Energy and Mineral Resources at the 11th Energy Working Group Meeting Indonesia-Netherlands. Netherlands: 18–20 September, 2006

Legowo E (2009) Biofuel development in Indonesia. Presentation by Ministry of Energy and Mineral Resources at the Research Workshop on Sustainable Biofuel Development. Jakarta, Indonesia: 4–5 February, 2009

Leow C (2008) Palm oil rises to record as Indonesian taxes may curb supply. Bloomberg.com. 5 Feb. http://www.bloomberg.com/apps/news?pid=20601087&sid=a4wxtklmOVB0&refer=home. Accessed 5 May 2008

Licht FO (2009) World ethanol & biofuels report vol.7 no.14. F.O. Licht, Ratzburg

Li G, Niu S, Ma L, Zhang X (2009) Assessment of environmental and economic costs of rural household energy consumption in Loess Hilly Region, Gansu Province, China. Renew Energy 34(6):1438–1444. doi:10.1016/j.renene.2008.10.018

MAFF (2009) Kokusan baio nenryo shinjidai [New era of domestically produced biofuels]. Presented at the Biofuels World 2009 Conference, 22 July, in Yokohama, Japan

Mang H-P (2008) Perspective of bioenergy and Jatropha in China. International consultation on pro-poor Jatropha development. IFAD, Rome

Martinot E, Junfeng L (2007) Powering China's development, the role of renewable energy. Worldwatch Institute, Washington, DC

Matsumoto N, Sano D (2011) Waste-based ethanol production and a sound material-cycle society: case studies on construction and food wastes in Japan. Institute for Global Environmental Strategies, Hayama

Matsumoto N, Sano D, Elder M (2009) Biofuel initiatives in Japan: strategies, policies, and future potential. Appl Energy 86(Supplement 1):S69–S76

Morimoto I (2008) Biomass utilization in China. 5th biomass-Asia workshop, Guangzhou, China

National Bureau of Statistics and NDRC (2007) China energy statistical yearbook 2007. China Statistics Press

NDRC (2007) Medium and long-term development plan for renewable energy in China (English draft). NDRC (National Development and Reform Commission), Beijing

NDRC (2008) Renewable energy development plan for the 11th five-year period (Japanese provisional translation). NDRC (National Development and Reform Commission), Beijing

OECD-FAO (2008) OECD-FAO agricultural outlook 2008–2017. Retrieved November, 2008 from http://www.oecd.org/dataoecd/54/15/40715381.pdf

Pan Jiahua, Li Meng, Wu Xiangyang, Wan Lishuang, Rebecca J. Elias, David G. Victor, Hisham Zerriffi, Chi Zhang, Peng Wuyuan (2006) Rural Electrification in China 1950-2004: Historical processes and key driving forces. Program on Energy and Sustainable Development Working Paper #60, December. http://pesd.fsi.stanford.edu/sites/default/files/WP_60%2C_Rural_Elec_China.pdf

People's Daily Online (2008) Sales of bioethanol-blended gasoline started in Guangxi (translated into Japanese). 21 April 2008. http://j.peopledaily.com.cn/2008/04/16/jp20080416_86876.html. Accessed 21 Apr 2008

PetroChina (2008) Largest biodiesel production from non-food grains (Chinese). Beijing
Pew Charitable Trusts (2010) G-20 clean energy factbook. Pew Charitable Trusts, Washington, DC
Pimentel D, Patzek T, Siegert F, Giampietro M, Haberl H (2007) Concerns over notes on biofuels in IPCC AR4 mitigation report and SPM. Geneva, 30 October 2007
Reuters (2007) Indonesia's Bakrie delays biodiesel plant to 2009. http://uk.reuters.com/article/2007/08/29/bakrie-indonesia-idUKJAK29769320070829. Accessed 13 May 2008
Romero J (2010) Fast-tracking renewable energy: pathways to sustainable energy utilisation. In: Sustainable consumption and production in the Asia-Pacific region: effective responses in a resource constrained world. IGES, Hayama, pp 197–214
Saigo M (2008) Nihon-gata baio neryo seisan kakudai ni mukete [Towards expansion of Japanese-style biofuel production]. Presented at the Biofuels World 2008 Conference, 9 July in Yokohama, Japan
Sakata I (2009) Toyota no baio nenryo heno tenbo [Toyota's perspective on Bio Fuels]. Presented at the Biofuels World 2009 Conference, in Yokohama, Japan
Sano D, Romero J, Elder M (2012) Jatropha production for biodiesel in Yunnan, China: implications for sustainability at the village level. Cambridge University Press, New York
Sun X (2007) Non-staple crops new sources for ethanol. China Daily. 14 June 2007
Takamizawa M (2009). Difficult steering of China's energy policies (translated from Japanese). Retrieved 6 April 2009, from http://news.searchina.ne.jp/disp.cgi?y=2009&d=0401&f=column_0401_001.shtml
UNEP (2008) Reforming energy subsidies: opportunities to contribute to the climate change agenda. United Nations Environment Programme, Department of Technology, Industry, and Economics, Paris
USAID (2009) Biofuels in Asia: an analysis of sustainability options. United States Agency for International Development, Washington, DC
USDA (2008) China bio-fuels annual 2008, GAIN report number CH8052. United States department of Agriculture, Washington, DC
Wani SP, Chander G (2012) Jatropha Curcas biodiesel, challenges and opportunities: is it a panacea for energy crisis, ecosystem service and rural livelihoods. In: Carels N, Sujatha M, Bahadur B (eds) Jatropha, challenges for a new energy crop. Springer, New York, pp 311–331. https://doi.org/10.1007/978-1-4614-4806-8_16
Zhang Y, Siang CC (2007) Alternative fuel implementation policy in China and its assessment. Institute of Energy Economics, Japan, Tokyo

Part IV
Impacts on Land Use and Ecosystem Services: Social, Economic and Political Impacts

Chapter 9
Social, Economic, and Political Impacts

Masahiro Matsuura and Hideaki Shiroyama

9.1 Introduction

Authors have conducted two case studies of stakeholder analysis on the utilization of sugarcane-based bioethanol in Brazil and palm-based biodiesel in Indonesia. Our research has focused primarily on the aspects of exporting these biofuels to Japan in order to give more concreteness to the stakeholder interviews. While these cases provide unique contexts in the production of feedstocks and distillation processes in each country, they also indicate common features that have to be considered in the policies, either at the international or the national level, for the sustainable utilization of biofuels.

9.2 Case from Brazil

9.2.1 Method

We have conducted interviews with a wide range of stakeholders in 2008. The interviewees are selected based on our literature review and inputs from the partner in Brazil (Table 9.1).

M. Matsuura (✉)
Graduate School of Governance Studies, Meiji University,
1-1 Kanda Surugadai, Chiyoda-ku, Tokyo 101-8301, Japan
e-mail: mmatsuura@meiji.ac.jp

H. Shiroyama
Graduate School of Law and Politics, The University of Tokyo,
Tokyo, Japan

K. Takeuchi et al. (eds.), *Biofuels and Sustainability*, Science for Sustainable Societies, https://doi.org/10.1007/978-4-431-54895-9_9

Table 9.1 Stakeholder interviewees in Brazil

Government	Industry	Research and civil society
Ministeiro da Agricultura	Uniao da Industria de Cana-de-Acucar (UNICA)	Universidade de Sao Paulo, Centro de Estudos Avancados em Economia Aplicada
Empresa Brasileira de Pesquisa Agropecuária (EMBRAPA)	Associacao Brasileira de Agribusiness (abag)	Economia & Energia (NGO)
Ministeiro de Ciencia e Tecnologia	PetroBras	ONG Vitae Civilis (NGO)
Banco Nacional de Desenvovimento Economico e Social (BNDES)	DEDINI	
	Mitsui company in Brazil	

9.2.2 *Key Stakeholders*

Based on the interview results, the following categories, described below, are identified as the key stakeholders that have interests in increasing the ethanol production for exports to Japan. Production of sugarcane in Brazil is concentrated mainly in the State of Sao Paolo and Brazil's northeastern region called Nordeste. The available land in the Nordeste region, however, is limited due to its hilly topography, and a large-scale production increase is unlikely. Therefore, we have focused on the possibilities of increased production in the State of Sao Paolo and other states on its north (namely, Mato Grosso do Sul, Minas Gerais, and Goias).

9.2.2.1 Industrial Sector

Sugarcane Plantation and Distillery Recently sugarcane plantations are mainly large scale and structured as a well-managed development project. Sugar mill and distillery are often developed as integrated part of the plantation. Because of large-scale investment requirement, they have concerns about the large fluctuation in demand and price and uncertainty in the investment environment (including government policy and infrastructure development). They are also affected by government regulations on the environment and labor. They are also interested in the electricity price because they benefit from selling the electricity generated through burning bagasse.

Investors and Trade Companies Several major corporations (oil, automobile, and trade) that have stakes in the agricultural and automobile industries have already made commitments to invest in sugarcane plantations. They share the concerns about the return from their investment, as the plantation owners do. These investors are also involved in the development of infrastructure, in addition to that of plantations. In particular, they are interested in the bioethanol pipeline from the inland (e.g., Goias) to the Port of Santos in Sao Paolo state. An oil pipeline is already built along the way, and no concern about land acquisition is reported.

Mill and Distillery Plant Developers The large increase in the demand for ethanol has provided economic benefit to a few plant developers that provide crucial machineries to sugar mills and distillery. They are now even trying to expand their business by exporting their machineries to other developing countries that intend to explore sugarcane-based bioethanol production.

9.2.2.2 Government

President's Office At the time of the interviews, the previous administration, led by President Lula da Silva, had promoted the development of biofuels, and incumbent President Dilma Rousseff was likely to follow up on the promotion of bioethanol in the same line.

Ministry of Agriculture The Ministry has been very active in promoting the export of poultry to Japan. Bioethanol is the next target product that the Ministry intends to promote to Japan as the agricultural commodity. It is also concerned about the environmental impacts of plantation expansion and has been preparing a national map of appropriate areas for sugarcane plantations.

EMBRAPA It is a part of the Ministry of Agriculture and has been taking the central role in the research and development of biofuels, including biodiesel, in Brazil. Biofuel section was established in 2006.

Ministry of Mines and Energy The Ministry oversees the quality of biofuels and standard setting activities.

BNDES The government's investment bank promotes sustainable plantation development by offering a lower interest rate for energy recovery plant that burns bagasse. It has, however, concerns over the return from such investment as well.

9.2.2.3 Civil Society

Environmental NGOs They have strong concerns about the land use impacts of biofuel-related land use change on Cerrado and rainforests. They are also concerned about the impact of open burning on adjacent lands.

Labor Unions They are concerned about the employment at plantations. Due to the ban on open burning (which was introduced with an intention of environmental protection), the manual laborers are exposed to a harsher working condition because hazardous insects and leafs could not be removed before harvesting. On the other hand, mechanized harvesting, which relieves workers from the unsafe condition, means less demand for manual labor.

9.2.3 Key Issues in Biofuel Production in Brazil

9.2.3.1 Economic Issues

Unstable Demand Because the number of flex-fuel vehicles on the road is increasing rapidly in Brazil, the demand for bioethanol has been expanding with certainty. On the other hand, the large fluctuation of crude oil price in the last few years has been transcended to the unstable demand for bioethanols at the global scale. Higher the stability of public policy in the European Union, the United States, and Japan is, higher the certainty of demand for bioethanol production in Brazil.

Investment Environment Regulations on, as well as uncertainties in, foreign investments to Brazil limit the expansion of biofuel production. Brazil's economic policy has been relatively stable in the last few years, compared to previous administrations, and thus the uncertainty regarding the foreign investment is lower than before. However, the legacy of unreliable management of national economy in the past is affecting the decision of foreign investors.

Grid Connection Electricity generation by burning bagasse is rapidly increasing. The electricity is supplied not only to the distillery machineries but also to the grid. The supply to the grid would increase by 2,745 MW from 2007 to 2012. If the feed-in-tariff and similar regulation that promote power generation by plantations, the expected return from investing in plantation development would increase.

Infrastructure Development Pipeline is crucial in expanding the sugarcane production from the state of Sao Paolo toward its north. Transport of ethanols in the special lorries on the highway is costly and would harm the profitability of such expansion. The pipeline development toward the Port of Santos is a crucial element in predicting the future of biofuel development in Brazil.

9.2.3.2 Societal Issues

Employment Harvesting sugarcane is traditionally labor intensive, and the seasonal migrant laborers from the northern part of Brazil have undertaken the role. On the other hand, mechanization at plantations has progressed. It means that the expansion of new plantations does not necessarily transcend to the increased demand for manual labor force. Following the mechanization, skilled laborers who can operate agricultural instruments are likely to benefit from new employment opportunities. Meanwhile, those unskilled manual laborers need to be supported by providing vocational education and other job opportunities.

Labor Safety The ban on open burning has problems with labor safety. Scorpions and snakes cannot be removed from the field before harvesting. The sharp edge of the sugarcane leaf is another kind of risk to the laborers. On the other hand, there is a political pressure to ban open burning from the perspective of environmental protection (including climate change and air pollution). The balance between these two pressures must be well balanced.

Income Disparity Within Brazil Sugarcane harvesting has traditionally been undertaken by the seasonal migrant workers from the north where the economic development is much slower than in it is other region. The majority of sugarcane plantation developments in Brazil is expected in the southern part of Brazil and this trend may even widen the income disparity between these regions. Meanwhile, the Brazilian government is promoting feedstock production for biodiesel in the northern Brazil, which have implications on adjusting the income disparity.

9.2.3.3 Environmental Issue

Land Use Environmental NGOs have strong concerns about impacts on the environment. In addition to those on rainforests, they argue that uncontrolled developments in the Cerrado area could pose a serious threat to the environmental benefits from the area. Land use impacts of sugarcane production are addressed by other contributors to this book by using life cycle assessment and other scientific methods. Even those who promote the expansion of sugarcane plantation admit the existence of such concerns about the environment. On the other hand, they argue that the physical area for sugarcane plantation is relatively small compared to the existing pasture and underused land. They also argue that the impact could be controlled through the mapping effort by the federal government and land use regulations by the state government. The tension between developers and environmental advocates needs to be addressed by stakeholder processes that incorporate sound science and forecasts.

Open Burning As we have mentioned before, open burning is another important environmental issue. The ban on open burning leads to the mechanization of harvesting. Therefore, this issue must be considered in connection with other societal issues such as national labor policy.

9.2.3.4 Political issue

Biofuel Deployment Policy of the Importing Countries Export demand is a political issue as well. For instance, several environmental NGOs challenge the prospect of CO2 reduction through sugarcane-based bioethanol fuels. Such pressures from the civil society organizations increase the political uncertainty of biofuel deployment policies in each country and region. Even there is a wide gap between the Japanese government's stated goals and the actual deployment of biofuels. From the perspective of Brazil as an exporting country, policies of other importing nations and their implementation are a crucial issue in forecasting the export demand for bioethanol.

In order to exert influence in these target nations, Brazilian government has been active in promoting bioethanol through conferences, such as "Biofuels as a Driving Force of Sustainable Development" in November 2008, and other kinds of bilateral discussions.

Alliance with Other Feedstock-Producing Nations Brazil has been active in exploring strategic relationships with other nations in the field of bioethanol production, and partnering with other Latin American and African nations for bioethanol production is likely. In the very long run, these bioethanol suppliers could form a cartel, something like a bioethanol counterpart for the OPEC.

Stakeholder Dialogue In Brazil, the Sugarcane Discussion Group was established for facilitating the collaboration among stakeholders. It is convened by the sugarcane producer's organization (UNICA) and hosts dialogues for environmental NGOs and labor associations to discuss on open burning and other relevant issues.

9.3 Case from Indonesia

9.3.1 Method

We carried out interview surveys of various stakeholders in Indonesia in September 2009. Prior to the survey, we identified the interviewees from relevant literatures as well as by consultation with the research collaborators inside and outside Japan.

9.3.2 Key Stakeholders

Based on the interview survey, we identified the following organizations as the major stakeholders in increasing biodiesel production with the primary purpose of exporting to Japan. Both palm and *Jatropha* were examined in respect of the biodiesel production. Given the scale of the procurement of raw material, however, the palm oil-based biodiesel production appeared realistic in the short term. Accordingly, we focused our survey on palm oil and identified the following stakeholders:

9.3.2.1 Industrial Sector

Palm Producers They are generally divided into private large-scale farms (plantations), small-scale farms, and government-owned farms. The plantations carry out their operations from the plantation development through to the product development on their own. The area planted by small-scale farms is expanding at a considerable rate of 25% per year. Control of independent farms, however, is difficult.

Biodiesel Producers The APROBI is an organization of the producers. The current membership stands at 22 companies.

Financial Institutions The development of plantations requires investment in which Japanese financial institutions are said to be involved. There is also a strong demand from NGOs for sustainable investment.

Trading Companies Major Japanese trading companies are involved in the palm oil production and trading and said to be interested in biofuel business as well. At present, Nippon Biodiesel Fuel Co., Ltd. and others are carrying out small-scale export to Japan on an experimental basis.

9.3.2.2 Government

Central Government In addition to strong sectionalism among ministries and the lack of a mechanism for them to coordinate with one another, the complexity of the jurisdictional coverage creates policies governing the processes from palm production to biofuel production that are not coordinated.

Ministry of Agriculture They formulate development plans and promote their policy from the perspectives of farm development.

Ministry of Forestry They are in charge of establishing a national policy on the conversion of forest to farmland. They also maintain statistics relating to forest.

Ministry of Energy They implement their policy on subsidies for energy which includes biofuel.

Ministry of Industry They have jurisdiction over the biofuel refining process.

Ministry of Environment The have jurisdiction over environmental assessment programs relating to farm development for biofuel crop cultivation, as well as the authority to regulate the quality of water discharged from the biofuel production processes and implement measures against global warming (which relate to palm plantation development in the context of discussions on granting credits for forest conservation).

Science and Technology Agency (BPPT) They have jurisdiction over the policies on science and technology relating to the biofuel development.

Local Governments Local governments were given the authority to issue permits for land use conversion as a result of the decentralization of government in 1999. Indonesia has 27 provincial governments under which regencies and cities operate. The regencies and cities are divided into sub-districts. Regulations vary from region to region as a result of the decentralization. The capacity of executive officers and corruption are problems.

9.3.2.3 Civil Society

Roundtable on Sustainable Palm Oil (RSPO) This is an organization of businesses and NGOs. It is working to build a consensus on the standards of sustainability as part of an approach to promote sustainable production and distribution of palm oil. The RSPO is headquartered in Malaysia with a liaison office in Indonesia. It has published the interpretation of the standards for Indonesia.

NGOs International and local NGOs are operating in Indonesia for the purpose of the forest conservation and the protection of indigenous peoples. They are active not only in Indonesia but also in consumer countries, such as the United States and EU. Some NGOs regard compliance with the RSPO standards as satisfactory, while others demand tougher measures.

Indigenous People The development may provide improvements in opportunities for employment and education for them. There is, however, a risk of these people losing a sustainable infrastructure for their livelihood.

Indonesian Consumers The price of crude palm oil temporarily spiked during the period from 2007 to 2008 corresponding to the rise in crude oil prices. The consumption of palm oil as food may be affected.

9.3.3 Major Issues Relating to Biofuel Production

The key issue which the Indonesian palm industry faces is the sustainability of the production process, including its impact on the environment and society.

Biodiversity may be lost and the rights of the indigenous peoples violated as the result of deforestation and peat bog development as part of new plantation development as well as the destruction of the existing ecosystems due to these activities. If small farmers practice their slash-and-burn method of farming in the process of switching to palm cultivation, peat bogs, which are extremely flammable during the dry season, can be easily ignited and develop into forest fires. Such forest fires will inevitably lead to the loss of biodiversity and the violation of the rights of the indigenous people, as well as the emission of greenhouse gases (GHG).

Water pollution in these regions as a result of wastewater discharged from the oil extraction process as well as the use of agricultural chemicals and fertilizers during cultivation may also lead to the loss of biodiversity and the violation of the rights of the indigenous people. It is apparent that the ecology of wildlife and the livelihood of the indigenous peoples will be in a very vulnerable situation as the logical conclusion of these various issues.

In addition to the abovementioned problems which are applicable to the palm industry as a whole, particular important issues relating to the production of biodiesel from palm oil include a macroscopic problem of an increased demand for biofuel competing with the demand for food supply and a question as to whether replacing petroleum-based fuels by biodiesel actually leads to the reduction of the GHG.

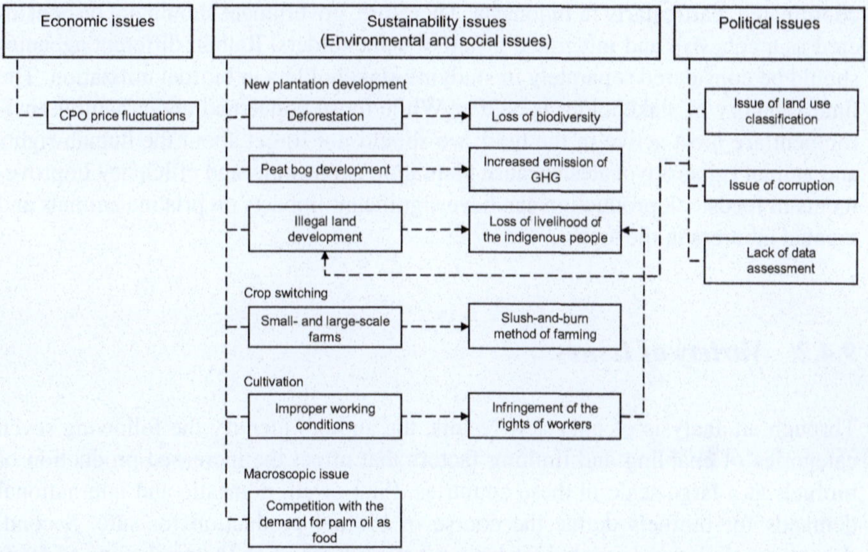

Fig. 9.1 Key issues in promoting palm oil biofuel in Indonesia

The issues relating to the palm oil industry and the palm oil-based biofuel production are intertwined in an extremely complex way. An overview, however, indicates that these problems can be divided into environmental and social issues, which may arise from palm oil farming and biofuel production (i.e., an issue of sustainability of the palm production); political issues, which prompt environmental and social issues; and economic issues, which may become an obstacle for the palm oil-based biofuel production in the future. Figure 9.1 illustrates the major issues.

9.4 Lessons from Two Cases

9.4.1 *Variety of Stakeholders*

There are six categories of common stakeholders in the production of biofuels in these countries. The first category is feedstock producers, including plantations, smallholders, and manual laborers. The second category is refineries, including those companies operating biofuel refineries as well as those producing machineries and developing plants. Investors are the third category. It includes trade companies and banks (national/corporate). The fourth category, transportation operators, is often neglected in the study of biofuels. Utilizing biofuels at the global scale means that they will be an equivalent to crude oil, which means safe transportation of biofuels will be a major concern from the perspective of protection of ocean environments as well as of geopolitics. The fifth category is government agencies. Multiple ministries and agencies are involved even in a single country, and they are often ill coordinated, particularly in Indonesia. Therefore, government should not be considered as a cohesive and integrated group of stakeholders. Rather, different agencies should be considered separately in studying stakeholders in biofuel utilization. The final category of stakeholder is NGOs. While those concerned about natural environment are most active in the field, we should not forget about the human rights and animal rights advocates because plantation expansions and efficiency improvements in feedstock production can have significant impacts on pristine animals and manual laborers in the area.

9.4.2 Variety of Issues

Through an analysis of interview results, the authors identify the following seven categories of enabling and limiting factors that affect the increased production of biofuels at a large scale in these countries. First of all, domestic and international demands for biofuels define the course of biofuel production for sure. Second, domestic policies and regulations have substantial effect on the production of feedstock and the deployment of biofuels in these nations. Third, domestic political culture also has impacts on it. This is different from policies itself; it is about how policies change and how well they are implemented. Fourth, pressures from supranational institutions and international NGOs can have an influence from the demand site. Fifth, availability of transportation infrastructure is a crucial factor in the supply chain of feedstock and biofuels. For instance, further development of sugarcane-based ethanol is likely to require the development of pipelines. Sixth, investment environment in each country has influences on the future development. While foreign investors have substantial interest in the development, they are also concerned about the uncertainties associated with the investment in these countries particularly because of the legacy of instability of these nations' economic policy. Seventh, mechanization of harvesting processes is another crucial factor in the production of feedstock. And finally, R&D for the improvement of feedstock species is likely to be a major challenge in a near future. In order to achieve sustainable development, plantation owners are now faced with the challenge of increasing productivity per unit of area. Improving the yield, as well as increasing the portion of useful content in each crop, through hybridization and genetic engineering will be required for the sustainable future.

9.4.3 Implication to the Japanese Government's Policy on the Importation of Biofuels

Through interviews with stakeholders in Brazil and Indonesia, the authors identified a few areas that the Japanese government can contribute for the sustainable utilization of biofuels.

First of all, the lack of government's committed mandate and/or deployment strategy is causing confusion on the side of potential exporters. For instance, Japanese government has expressed its interest in importing palm-based biodiesels from Indonesia particularly during the visit of Prime Minister and other senior officials to Jakarta. Such informal encouragements had certain influence among the biodiesel producers. On the other hand, Japan has not yet imported a sizable amount of biodiesel from Indonesia. A few stakeholders, including those in Brazil, explicitly expressed their frustration with the lack of more formal commitment on the Japanese side. Therefore, it would be beneficial for all stakeholders in this field if the Japanese government sets official target regarding the import of biofuels.

Secondly, as mentioned above, the lack of appropriate transportation infrastructures in these countries could be a major bottleneck for importation. This is an area in which the Japanese government can assist through developmental aid and other schemes to fund infrastructure development projects. The Japanese government and aid agencies can be strategic in selecting the applications by considering proposed projects' implication to the transport of biofuel to Japan and other parts of the world.

The last concern about the Japanese government is its sustainability standard. While it is an area where scholars and stakeholders are still debating about, one pragmatic proposal would be to internalize sustainability standards, such as the one proposed by Roundtable on Sustainable Palm Oil, into Japanese regulatory structure. While some NGOs are still critical about the utilization of biofuels at the global scale, such a strategy could justify the biodiesel import to some extent because these international efforts have consciously involved nongovernmental stakeholders and tried to seek an agreement that these concerned parties could live with.

Chapter 10
Stakeholder Perceptions of the Ecosystem Services and Human Well-Being Impacts of Palm Oil Biofuels in Indonesia and Malaysia

Raquel Moreno-Peñaranda, Alexandros Gasparatos, Per Stromberg, Aki Suwa, and Jose A. Puppim de Oliveira

10.1 Introduction

Palm oil is an essential part of the diet of many people around the world. It is also a key raw material for the processed food, and pharmaceutical and cosmetics industry. According to the Food and Agricultural Organization of the United Nations (FAO), palm oil is the most widely produced vegetable oil accounting for 33% of the global vegetable oil production in 2014 (FAO 2017). At the same time, palm oil production has increased more than any other type of vegetable oil since 1961, recording a staggering 39-fold increase. The FAO projects that the production of oil crops will almost triple (from 100 million tons to 293 million tons) in 2050 with a large part of this increase being due to oil palm expansion.

The original version of this chapter was revised: incorrect citation and its reference were removed. A correction to this chapter can be found at https://doi.org/10.1007/978-4-431-54895-9_16

R. Moreno-Peñaranda (✉)
College of Arts and Sciences, The University of Tokyo, Tokyo, Japan
e-mail: raquelmp@gmail.com

A. Gasparatos
Integrated Research System for Sustainability Science (IR3S), The University of Tokyo, Tokyo, Japan

P. Stromberg
United Nations University – Institute of Advanced Studies, Yokohama, Japan

Environmental Economics Unit, Swedish Environmental Protection Agency, Stockholm, Sweden

A. Suwa
Faculty for the Study of Contemporary Society, Kyoto Women's University, Kyoto, Japan

J.A. Puppim de Oliveira
Fundação Getulio Vargas (FGV), São Paulo School of Business Administration (FGV/EAESP) and Brazilian School of Public and Business Administration (FGV/EBAPE), São Paulo, Brazil

K. Takeuchi et al. (eds.), *Biofuels and Sustainability*, Science for Sustainable Societies, https://doi.org/10.1007/978-4-431-54895-9_10

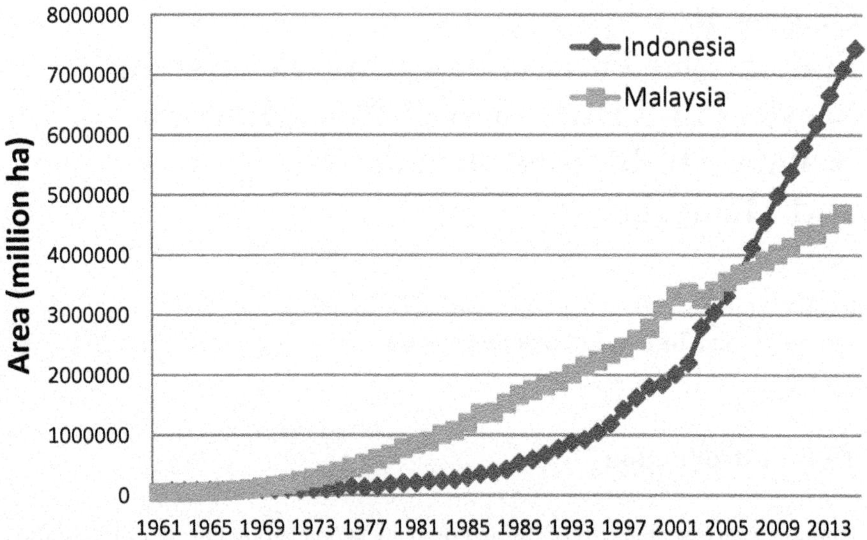

Fig. 10.1 Area under oil palm cultivation in Indonesia and Malaysia, 1961–2014 (Source FAO 2017)

Globally, most of the palm oil production is located in Southeast Asia and particularly in Indonesia and Malaysia. The two countries accounted for 85.4% of global production in 2014 and for 86.0% of palm oil exports in 2013 (FAO 2017).[1] In fact, the huge increase in palm oil production globally has been spearheaded by the oil palm expansion and yield increases in the two countries. Between 1961 and 2014, palm fruit yield increased by 27.7% in Indonesia and by 77.4% in Malaysia (FAO 2017). However, these yield increases are nowhere near to the expansion of the area under oil palm cultivation, a 106-fold expansion in Indonesia and a 108-fold expansion in Malaysia (Fig. 10.1).

There is evidence to suggest that oil palm production mainly happened at the expense of forested land rather than land already under agriculture (Gibbs et al. 2010). According to Koh and Wilcove (2008), 55–59% of oil palm expansion in Malaysia and at least 56% in Indonesia occurred at the expense of primary forests. Estimates suggest that since 1990, 1 million ha and 1.7–3 million ha of forest were lost in Malaysia and Indonesia, respectively, due to oil palm (Fitzherbert et al. 2008; Koh and Wilcove 2008). There have been fears that the lack of careful planning in the ongoing expansion of oil palm in Indonesia might lead to significant additional deforestation (Koh and Ghazoul 2010).

Smallholders play a significant role is the palm oil sector in Indonesia (Pricewaterhouse Coopers 2010; Gatto et al. 2015). Most smallholders have entered the palm oil sector through state-sponsored agreements between private companies

[1] In 2013, palm oil was the seventh most widely traded agricultural commodity in monetary terms at USD 33.8 billion with the biggest importers being India, China, the Netherlands, Pakistan, and Nigeria (FAO 2017).

Table 10.1 Drivers of biofuel policies in Indonesia and Malaysia

	Energy	Economy			Society		Environment	
	Energy security	Trade balance	Price of petroleum	Improve economy	Increase agricultural employment	Improve rural incomes	Climate change	Air quality
Indonesia	√	–	–	√	√	√	–	–
Malaysia	–	√	√	√	–	–	√	–

Source: Zhou and Thomson (2009)

and local communities (plasma schemes).[2] There have been two main types of plasma schemes, the PIR Trans schemes (nucleus-community plantation), which target individual farmers, and the KKPA schemes, which target cooperative farmers. Lately there have other types of land-leasing agreements such as the kemitraan schemes and the plantation revitalization schemes (McCarthy et al. 2012a).

Despite some differences, these schemes stipulate that the palm fruit produced in the land "leased" by local communities has to be sold to the private company with which the local community has made the arrangement (McCarthy 2010; McCarthy et al. 2012a). In return, the company provides all necessary inputs and knowledge for oil palm cultivation. Such schemes determine to a large extent the productivity targets, and as an extension the production practices employed (e.g., fertilizer use). As a result these plasma agreements can be viewed as a major determinant of the local environmental and socioeconomics impacts of oil palm in Indonesia.

Currently palm oil is mostly absorbed by the food industry, a trend that is likely to continue in the short-to-medium term.[3] However in the late 2000s, palm oil was considered as a particularly promising feedstock for biodiesel production within the two countries and abroad. For different reasons, Malaysia and Indonesia took steps to promote the use of biofuels, mainly palm oil biodiesel, (Table 10.1). In 2010 Malaysia had already mandated the use of B5 (diesel containing 5% biodiesel), but Indonesia had not yet enacted mandatory biofuel blending (REN21 2010). Currently both countries have enacted relatively ambitious biodiesel mandates: B10 in Malaysia and B20 in Indonesia (REN21 2016).

At the international level, the 2009 European Union Directive on Renewable Energy (EU-RED) required that by 2020 10% of all transport fuel used within the EU should come from renewable sources (EU 2009). At that moment the EU was the largest producer and consumer of biodiesel (IEA 2011), so there were expectations of large increases of palm oil imports from Indonesia and Malaysia for

---

[2]There were also certain examples of oil palm expansion through the direct influence of the Indonesian government. For example the government of Aceh Barat Daya ruled that community members could establish oil palm plots (up to 2 ha per household) with the government providing necessary inputs such as fertilizer. This regulation led to massive land conversion by smallholder farmers (Tata et al. 2010), but the ensuing global financial crisis resulted in many Indonesian producers reducing or halting their planting program (NBPOL 2009).

[3]It is expected that in 2050, 42% of the global vegetable oil production will be diverted to industrial uses, when compared to 16% in 1976 and 24% in 2006 (FAO 2006).

conversion into biodiesel (Obidzinski et al. 2012). TE (2016) estimates that 45% of the palm oil used in the EU is used for biodiesel and 16% for electricity/heating (up from 8% and 14% in 2010). An immediate result of the above is that the palm oil industry became closely linked to energy markets, at least conceptually.

This perceived interest in palm oil-based biofuels in the late 2000s seems to have led to a sharp increase in the overall volume of imported palm oil in the EU (FEDIOL 2016). At the same time it contributed to the immense pressure from the environmental community for the adoption of more sustainable practices for palm oil production. This was part of the series of criticisms that the palm oil sector started facing since the late 1990s regarding its environmental and social performance (Schouten and Glasbergen 2011). Common practices in palm oil cultivation such as extensive monocultures, land clearing through fire, and pollution due to high fertilizer/pesticide use and palm oil mill effluent (POME) were identified as particularly damaging to the environment and biodiversity (Sect. 10.3).

Such concerns catalyzed the establishment of the Roundtable on Sustainable Palm Oil (RSPO) in 2004, a multi-stakeholder alliance promoting the sustainable production and consumption of palm oil. RSPO brings together important actors across the palm oil chain, such as large producers, smallholders, processors, traders, nongovernmental organizations (NGOs), and certifiers among others. Despite the initially slow take-up of the RSPO standard, the RSPO-certified growers accounted for approximately 28% of global palm oil production in 2012, with 42% of their product being already certified (RSPO 2012). However, the broader legitimacy of the RSPO has been questioned (Schouten and Glasbergen 2011; Partzsch 2009; Silva-Castaneda 2012; von Geibler 2013; Nikoloyuk et al. 2010). This is possibly because RSPO stakeholders have joined (and operate within) the alliance having radically different agendas and motivations (Schouten and Glasbergen 2011). Such divergent approaches within, and beyond RSPO, throw into doubt its ability to effectively promote the sustainable production and consumption of palm oil (Partzsch 2009).

In any case this growing connection between palm oil and biofuels in the late 2000s sparked some visible changes in the palm oil production chain as even companies that were not involved in biofuels (having produced palm oil for more than 100 years) were under increasing pressure to adapt their business practices to satisfy the environmental concerns of their clients.

Considering the above, this chapter elicits some of the concerns that key stakeholders involved in the RSPO process articulated during the early 2010s about the possible impacts of biofuel-driven palm oil expansion on the environment and human well-being. We use an ecosystem services perspective as a means of synthesizing the literature and highlighting the interlinkages between biofuel-driven ecosystem change and human well-being (MA 2005a, b). Section 10.2 provides a brief literature review of the key impact of oil palm cultivation and palm oil biofuel production and use on ecosystem services and human well-being. Section 10.3 outlines the methodological approach used to elicit stakeholder perceptions.

Section 10.4 compares the perceptions of different stakeholder groups involved in the RSPO process on how the palm oil industry affects local ecosystem services and human well-being (Sect. 10.4), while Sect. 10.5 outlines the main lessons learnt during this stakeholder engagement process.

10.2 Impacts of Palm Oil Biofuels on Ecosystem Services and Human Well-Being

10.2.1 Linking Biofuels and Ecosystem Services

The ecosystem services approach explicitly links ecosystem change to human well-being (MA 2005a, b). These are two key components of the biofuel debate that are evoked by proponents and critics of biofuels alike (Gasparatos et al. 2011). Studies have adapted the ecosystem services approach to synthesize knowledge about different biofuel value chains (e.g., Gasparatos et al. 2011) as well as assess specific impacts (e.g., Gissi et al. 2016; Romeu-Dalmau et al. 2017; Meyer et al. 2015).

Sections 10.2.2 and 10.2.3 collect and discuss the key environmental and socioeconomic impacts associated with palm oil biodiesel production/use in Malaysia and Indonesia employing the conceptual framework popularized by the Millennium Ecosystem Assessment (MA 2005a, b) as it has been adapted for biofuels by Gasparatos et al. (2011).

Recent studies have shown that biofuels can be major agents of ecosystem change due to land use and cover change (LUCC), pollution, climate change, introduction of alien invasive species, and overexploitation (Gasparatos et al. 2017). Following the MA vocabulary, we collectively refer to these factors as the direct drivers of biofuel-induced ecosystem change. Consequently the drivers of biofuel expansion itself (i.e., energy security, climate change mitigation, rural development) mentioned in Table 10.1 can be seen as the indirect drivers of biofuel-induced ecosystem change (Fig. 10.2).

It should be noted that in the case of ecosystem services, the way the evidence is reported in the academic literature coincides to an extent with the typology of ecosystem services used in the MA conceptual framework. However with the exception of "health," the human well-being impacts of biofuels are not reported following the constituents of human well-being outlined in the MA framework (Gasparatos et al. 2011). Furthermore in the case of palm oil biodiesel, the constituents of human well-being seem to be interlinked. In order to overcome these challenges, we identify the main impacts of biofuels on human well-being as reported in the academic literature, i.e., rural development, energy security/access to energy, food security/access to food, and health and land tenure. We then proceed in each of these sections

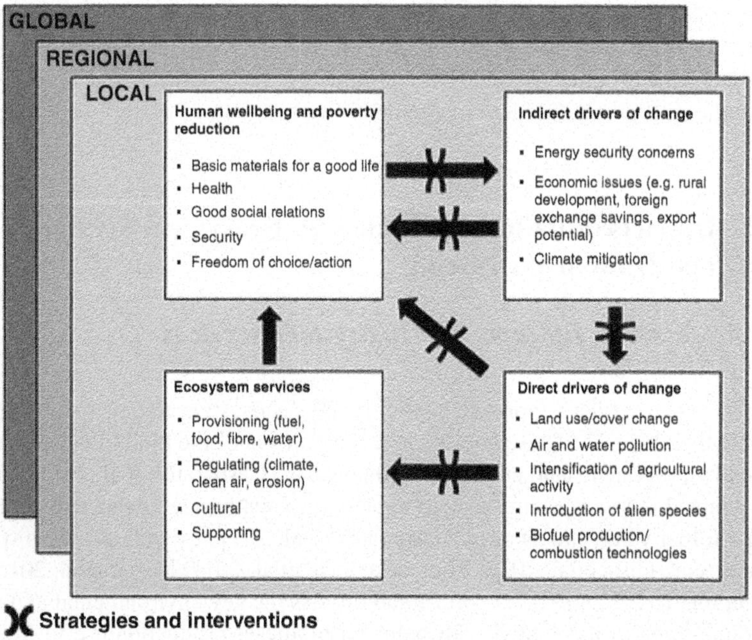

Fig. 10.2 MA conceptual framework related for palm oil diesel production/use (Gasparatos et al. 2011)

to discuss which of the MA constituents of human well-being are directly[4] impacted and through which mechanisms.

Figure 10.3 below depicts the linkages between ecosystem services and human well-being in the context of biofuel production. For example, biofuel expansion may increase access to fuel but also reduce access to food, hence affecting both security and basic materials supporting livelihoods. Lastly, strategies and interventions such as land use planning can enhance the ecosystem and social benefits resulting from the linkages between the four squares of Fig. 10.1. Examples of such response measures are given in Sect. 10.6.

10.2.2 Impact on Ecosystem Services

10.2.2.1 Feedstock for Fuel (Provisioning Service)

Oil palm fruits are the main ecosystem service provided by areas converted for oil palm production (Dislich et al. 2017). The palm oil derived from processing these fruits can be used for the production of biodiesel through trans-esterification

[4] Each of these issues affects indirectly all of the constituents of human well-being in the MA conceptual framework.

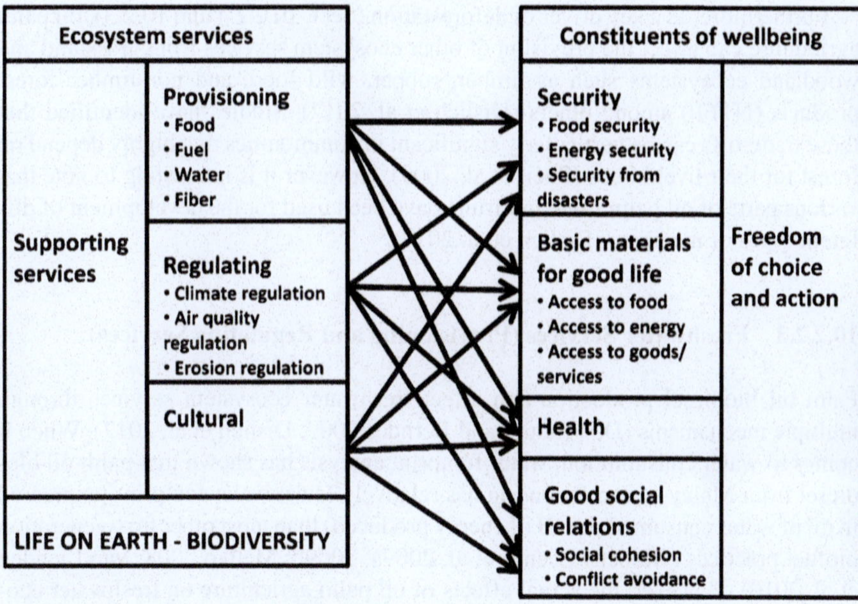

Fig. 10.3 Linkages between palm oil biodiesel production/use, ecosystem services, and human well-being (Gasparatos et al. 2011)

(Mekhilef et al. 2011). Oil palm agriculture can be highly productive, but at the same time it depends on the agricultural practices adopted (Sheil et al. 2009; Woittiez et al. 2017). Several studies have confirmed the large biodiesel potential from oil palm in Indonesia and Malaysia (e.g., Mukherjee and Sovacool 2014), but there is a need for more rational allocation of land resources to meet multiple objectives related to food, bioenergy, and biodiversity conservation (Harahap et al. 2017).

10.2.2.2 Food Crops and Woodland/Grassland Products (Provisioning Services)

As mentioned in Sect. 10.1, palm oil is the most widely produced vegetable oil globally, being a major component of the food industry. Biofuel feedstock production can sometimes entail the direct diversion of crops from food-related uses, potentially contributing to, among others, reduced local food availability and increases in food prices (Gasparatos et al. 2011; Schoneveld 2010) (Sect. 10.2.4.3). Mekhilef et al. (2011) report that close to 40% of Malaysian palm oil had been allocated for fuel production, putting pressure on remaining amount for vegetable oil demand.

The direct and indirect LUCC effects of oil palm expansion may affect local food production (particularly rice cultivation) either due to the direct loss of arable land or reduced water availability for agriculture (e.g., Oosterveer et al. 2014).

Furthermore, as a key driver of deforestation (Sect. 10.2.2.4 and 10.2.3), oil palm agriculture can affect the provision of other ecosystem services from grassland and woodland ecosystems such as timber, rubber, wild food, and non-timber forest products (NTFP) among others (Dislich et al. 2017). Studies have identified that these trade-offs can be particularly significant in communities that highly depend on forest for their livelihoods (Sheil et al. 2009). However it is interesting to note that various parts of oil palm trees and fruits have been used for the development of different types of medicine (Dislich et al. 2017).

10.2.2.3 Freshwater Services (Provisioning and Regulating Services)

Palm oil biodiesel production can affect freshwater ecosystem services through multiple mechanisms (De Fraiture and Berndes 2009; Dislich et al. 2017). When it comes to water consumption, water footprint analysis has shown that palm oil biodiesel from Malaysia and Indonesia has relatively lower water footprint (expressed in m^3 of water consumed per GJ of energy produced) than most other first-generation biofuel practices (Gerbens-Leenes et al. 2009a, 2009b; Mellko 2008; Van Lienden et al. 2010). However, the actual effects of oil palm agriculture on freshwater ecosystem services can be much more complicated as the conversion of forested land to oil palm monocultures can affect a series of functions related to hydrological cycles (Dislich et al. 2017). A recent meta-analysis of the literature suggests mostly negative effects due to decreases in a series of functions such as water storage, infiltration rate, regularity of supply, regulation of peak flows, water quality, and flood and drought prevention (Dislich et al. 2017).

When it comes to water quality, oil palm plantations are very fertilizer intensive in both countries (FAO 2004; FIAM 2009; FAO 2005). Fertilizer and pesticide residues can enter water bodies and potentially disrupt ecosystem functioning and negatively affect human health (refer to Sect. 10.2.4). The palm oil industry has also been identified as a major source of water pollution in Malaysia (Muyibi et al. 2008). Palm oil mill effluent (POME) is characterized by high levels of BOD[5] with approximately 2.5–3 tons of POME being produced for each ton of palm oil (Wu et al. 2010). However it has been suggested that POME can be used for oil palm but the environmental co-benefits of such practices are debatable.

10.2.2.4 Climate Regulation (Regulating Service)

Biofuels have been identified as potential climate mitigation options (e.g., IPCC 2007). Even though biofuel production/use can emit significant amounts of GHGs during their whole life cycle (Hess et al. 2009), several LCAs have shown that some biofuel practices can emit less GHG than fossil fuels during their whole life cycle. Palm oil biodiesel can provide significant carbon savings (up to 80%) when

[5] POME has BOD of 21,500–24,500 mg/L which is several times higher than that of sewage water.

compared to conventional fossil fuels (Menichetti and Otto 2009). Smeets et al. (2008) calculate robust GHG reduction potential of up to 75%. RFA (2008) reports a 46% GHG savings for palm oil biodiesel in Malaysia.

However most new oil palm plantations have been established on previously forested areas and often on former peatland forests (Carlson et al. 2012). Such LUCC effects can result in high carbon debts (Carlson et al. 2012; Koh et al. 2011; Moore et al. 2013; Ramdani and Hino 2013; van Straaten et al. 2015; Dislich et al. 2017) that might take several decades or centuries to repay. Danielsen et al. (2009) calculated that depending on the forest clearing method used, it would take 75–93 years for an oil palm plantation to compensate the carbon lost during the conversion of the initial forest, 600 years if that happens on peatland, and approximately 10 years if that happens on grassland. Fargione et al. (2008) report that the time to repay the biodiesel carbon debt would be 86 years if palm oil is established on forested land and 423 years if that forest is located on peatland. RFA (2008) calculates carbon payback time of 0–11 years for biodiesel from oil palm grown on grassland and 18–38 years on forested land.

10.2.2.5 Air Quality Regulation (Regulating Service)

Palm trees, like all other plants, emit volatile organic compounds (VOCs) and isoprene in particular. Hewitt et al. (2009) and Fowler et al. (2011) have shown that indeed VOC and nitrogen oxides (NOx) emissions, which are tropospheric ozone precursors (O_3), are greater from oil palm plantations than from primary rainforest. Sometimes the land that is used for oil palm production is cleared through the use of fire (e.g., Van der Werf et al. 2008). Biomass burning has been identified as major sources of atmospheric pollution and GHG emissions, affecting significantly atmospheric chemistry and biogeochemical cycles among other impacts (Crutzen and Andreae 1990). Communities adjacent to oil palm plantations often report declining air quality due to activities within the plantations (Obidzinski et al. 2012).

10.2.2.6 Erosion Control (Regulating Service)

Mature oil palm plantations in Malaysia have a soil erosion rate of approximately 7.7–14 tons/ha/year with erosion rates being even larger during the early years of the plantation when a complete palm canopy has not yet been established (Stromberg et al. 2010; Lee et al. 2012). In order of decreasing soil erosion hazard,[6] de Vries et al. (2010) ranked the most commonly used feedstocks as follows: cassava, soybean, sugarcane, sorghum, corn, sugar beet, winter wheat, oil palm, and winter rapeseed.

[6] This is an indicative ranking that can depend on the characteristics of the soil itself and the cultivation method adopted among other factors.

However when compared to natural ecosystems, oil palm plantations have much lower erosion control potential (Dislich et al. 2017; Buschman et al. 2012). In some cases eroded soil can enter water bodies further deteriorating local water quality (Obidzinski et al. 2012) or can result in the loss of soil organic carbon further contributing to the loss of climate mitigation services (Guillaume et al. 2015).

10.2.2.7 Cultural Services

For local communities and indigenous people, cultural services frequently form an important element of their culture and can be threatened by land use change, e.g., through habitat destruction and the displacement of traditional crops (MA 2005a). It has been suggested that biofuel-induced deforestation can affect indigenous people disproportionately. For example, almost half of Indonesia's population depends on ecosystem goods and services from forests with approximately 40 million of these people being indigenous and having been already affected (Tauli-Corpuz and Tamang 2007).

Cultural ecosystem services are a particularly understudied topic within the literature related to palm oil impacts. While these are some cultural benefits related to spiritual values in areas that oil palm grows naturally, the evidence suggests overwhelmingly negative impacts on cultural ecosystem services in areas that oil palm is grown intensively and has replaced forest (Dislich et al. 2017).

10.2.3 Impacts on Biodiversity

Biofuel production (particularly feedstock cultivation) can have multiple negative impacts to biodiversity (Gasparatos et al. 2017). Biodiversity is not an ecosystem service per se but "…the foundation of ecosystem services to which human well-being is intimately linked" (MA 2005: 18). According to the MA, there are six main direct drivers associated with biodiversity decline: habitat destruction, overexploitation, invasive species, disease, pollution, and climate change (MA 2005a). Palm oil biodiesel production/use can be strongly linked to at least three of these drivers, i.e., habitat destruction, pollution, and climate change with habitat destruction being considered as the most important. Overall several systematic reviews have highlighted the overall negative biodiversity outcomes of the conversion of natural habitats to oil palm plantations (Dislich et al. 2017; Savilaakso et al. 2014).

Oil palm cultivation in large-scale monocultures is by definition inhospitable to biodiversity. Oil palm plantations contain much fewer species than primary forests (e.g., Fitzherbert et al. 2008; Danielsen et al. 2009; Foster et al. 2011). Additionally several studies have found that the majority of the forest species was lost and replaced by smaller numbers of non-forest species with the subsequent animal communities being dominated by a few generalist species of low conservation value

(Danielsen et al. 2009). Not surprisingly, plant diversity within oil palm plantations was impoverished compared to forests due to regular maintenance and replanting (every 25–30 years) of oil palm fields (Fitzherbert et al. 2008; Danielsen et al. 2009).

10.2.4 Impacts on Human Well-Being

10.2.4.1 Rural Development

Indonesia and Malaysia currently have a large and highly competitive palm oil production sectors that are very important to their respective national economies. The oil palm sector can provide substantial employment and income opportunities (Rist et al. 2010; Cahyadi and Waibel 2013). Winrock (2009) estimated that up to 57% of Riau's population, and between 10–50% in 11 other Indonesian regions, were supported one way or another by the oil palm industry (including employees and family dependants in downstream processing and associated services).

Local communities in Indonesia often perceive oil palm cultivation as a promising livelihood activity (Rist et al. 2010). Oil palm production can have higher income returns to land and labor for smallholders, but the overall livelihood benefits can depend significantly within (and across) communities (Rist et al. 2010; McCarthy 2010) (see Box 10.1). For example, while income from oil palm production contributes significantly to the livelihoods of independent smallholder households (e.g., Lee et al. 2014; Budidarsono et al. 2012), it can vary depending on the agricultural practices adopted (Lee et al. 2014). In some cases the received income can be severely reduced after paying the initial loans that allow them to be involved in oil palm agriculture, but the repayment period can depend on multiple circumstances (Feintrenie et al. 2010).

According to legislation, wages for permanent plantation workers should be at least equal to the provincial minimum labor payments in Indonesia. While locals can complement their farm income through temporary work in plantations (Tata et al. 2010; McCarthy 2010), in some regions, plantation jobs are often monopolized by transmigrants (Obidzinski et al. 2012, 2014). It is also worth mentioning that working in oil palm plantations is often a strenuous activity, with, sometimes, low labor standards and substantial gender disparities (Li 2015).

Despite its potential to improve rural development, the oil palm sector operates in isolation in many Indonesian provinces and has limited economic multipliers (Obidzinski et al. 2014). Often this happens because employment benefits do not always reach the local communities, as permanent contract workers for agricultural labor and management in large plantations are usually transimmigrants (transmigrants) (Tata et al. 2010; Obidzinski et al. 2012; McCarthy 2010; Obidzinski et al. 2014).

It should be also noted that shifting to biofuel feedstock production can be a risky endeavor particularly for independent smallholders (Feintrenie et al. 2010). High market and production chain uncertainty as well as difficulty in complying with certain types of production standards can expose smallholders to the financial risk of not getting adequate returns on their investment or even being excluded altogether by the oil palm value chains (Jelsma et al. 2017; Cahyadi 2013). Price volatility in food commodities has been very prevalent since the 2000s, while adding the generally high volatile nature of energy markets in the equation, it can make decision regarding a shift toward biofuel feedstock production more difficult to handle particularly for smallholders (Woods 2006; Robles et al. 2009; DTE 2005).

Box 10.1: Local Income from Oil Palm Cultivation
Mulyoutami et al. (2010) studied oil palm adoption in Tripa (Aceh, Indonesia) as part of the 2004 tsunami rebuilding programs. They tracked government incentives, particularly smallholders, to switch from their previous economic activities to cultivate oil palm. They reported that smallholder plots in the Nagan Raya district had 120–150 oil palms per hectare and generated a gross production value of IDR 600,000–1,500,000 (approximately USD 67–168) per month per hectare. In 2010, the price of a fresh fruit bunch ranged between IDR 700,000 and 1,050,000 (approximately USD 80–110) per ton. A local survey in the Ladang Baru area showed that income for oil palm smallholders was higher than income from other local economic activities (e.g., 160,000–5,500,000 IDR/month compared with IDR 120,000–2,800,000 for oil palm plantation workers and IDR 115,000–750,000 for fishing). As mentioned above, wages for permanent plantation workers are regulated and should be at least the provincial minimum labor payments. In the case of Aceh province, this was IDR 1,000,000 per person per month in 2008. While such laborers tend to be immigrants, locals often take on sporadic day work to complement their farm income, earning a wage of IDR 25,000–40,000 (±USD 2.80–4.45) per day (Mulyoutami et al. 2010).

10.2.4.2 Energy Security and Access to Energy Resources

Several life cycle analyses (LCAs) have concluded that palm oil biodiesel production in Indonesia and Malaysia are net-energy providers, resulting in fossil energy savings of up to 80% (Zah, et al. 2008; Harsono et al. 2014; de Vries et al. 2010). A comparative LCA study ranked different biodiesel production chains according to their decreasing energy consumption, as follows: soybean (Argentina), soybean (Brazil), rapeseed (EU), rapeseed (Switzerland), palm oil (Malaysia), and soybeans (the USA) (Panichelli et al. 2009). Such findings suggest that palm oil biodiesel can be a feasible energy options in Malaysia and Indonesia, possibly enhancing national energy security.

In fact energy security has been identified as a major driver of biofuel production in Indonesia and Malaysia (Zhou and Thomson 2009). In spite of its significant domestic fossil fuel endowments, Indonesia is a net importer of crude oil and expects a strong increase in population. For these reasons there a strong need for alternative energy sources. The Presidential Decree No. 5 (2006) stated that biofuel would fulfill 5% of the total energy consumption by 2025 (APEC 2010). The Ministry of Energy and Mineral Resources stated that 520,000 tons of biodiesel were produced in 2007 in eight biodiesel plants, with another 15–17 planned for 2011, producing an additional 2 million kL of biodiesel (Zhou and Thomson 2009). The biofuel mixes B5 (Biosolar) and E5 (Biopertamax) have been available through the state-owned oil firm Pertamina. In January 2008, the rising international price of palm oil made Pertamina reduced the percentage of biofuels in its Biosolar and Biopertamax fuels from 5% to 2.5% (APEC 2010). Currently both countries have enacted relatively ambitious biodiesel mandates: B10 in Malaysia and B20 in Indonesia (REN21 2016). However the programs in both countries have been criticized about their effectiveness in boosting national energy security (Putrasari et al. 2016; Rahyla et al. 2017).

Oil palm waste such as empty fruit bunches can also be used as a feedstock for electricity generation and ethanol and/or biogas production having ripple positive effects on national and local energy security (e.g., Begum et al. 2013; Jinn et al. 2015). The Indonesian government committed to promote local energy security through the energy self-sufficient villages (ESSV) program. Bioenergy from oil palm was one of the energy sources considered, but to our best knowledge the program failed to produce good results, despite its vision to create the capability in thousands of villages to meet their own energy demand from locally available renewable resources such as biofuels, hydropower, and wind energy (IGES 2010).

10.2.4.3 Food Security and Access to Food

Biofuels can affect food security through multiple mechanisms related to the four pillars of food security proposed by the Food and Agricultural Organization (i.e., availability, access, utilization, and stability) (Wiggins et al. 2015). For example in Sect. 10.2.2.2, it was shown how feedstock production can divert food crop production, essentially reducing food availability through a trade-off between two provisioning ecosystem services (i.e., feedstock vs. food) (see also Sayer et al. 2012). At the same time, the income received from oil palm production and employment in plantations can be used to buy food, increasing thus household access to food and improving their nutrition (Euler et al. 2017). Furthermore smallholders can invest this income to buy agricultural inputs or have better access to inputs due to credit access as a result of their outgrower contracts functioning as collateral (Cramb and Sujang 2012; Cramb and McCarthy 2016). Environmental change may also affect local food production. For example, in areas of the Tripa province, the expansion of oil palm plantations led to water shortages in swamps, prohibiting the development of rice paddies (Mulyoutami et al. 2010).

Food security is a multidimensional concept, so effects such as those discussed above are not only difficult to be delineated at the local level but must be understood with respect to the national and international context. In Indonesia, the increased export demand for palm oil (Sect. 10.1) has possibly increased risks to palm oil shortages for Indonesian consumers. While palm oil prices can vary within year(s), they increased considerably in two very discrete spikes in the late 2007 (from USD 540/ton in early 2007 to USD 1440/ton in mid-2008) and mid-2010. This was possibly driven by demand for biofuel feedstock that also drove up the prices for sugar, grains, and vegetable oils (including palm oil), hence increasing households' living expenditures (OECD 2008) (Sect. 10.1). Such concerns have in several occasions prompted the Indonesia government to impose export taxes on palm oil, as a way to secure sufficient supply for domestic users. For example, in January 2011, the Indonesian Trade Ministry announced an increase of the tax to 25 percent to a large extent to avoid escalating food prices (Bouët and Laborde Debucquet 2016). While subsistence farmers may not be directly affected by changes in international commodity prices, poor people in food-deficit developing nations are considered as particularly vulnerable considering that they use a very large fraction of their income on food (e.g., Runge and Senauer 2008).

10.2.4.4 Health

Gasparatos et al. (2011) report several cases around the world where individual and public health have been compromised due to biofuel expansion. Health threats can be due to labor conditions in plantations (e.g., due to strenuous work), agricultural practices employed (e.g., use of agrochemicals, land clearing through fire), and malnutrition as a result of rising food prices. Several publications have reviewed the health hazards of working and living in the vicinity of oil palm plantations (e.g., Ng et al. 2014; UNICEF 2016). Regarding malnutrition, while there is evidence to suggest lower malnutrition levels in villages involved in oil palm activities (e.g., Budidarsono et al. 2012; Euler et al. 2017), there have been recorded instances in Indonesia of mothers in poor families lowering their food intake in order to feed their children when food prices rose (partly due to biofuels), (Actionaid 2010). Furthermore the air pollution health effects that have resulted from forest fires, to an extent for land clearing for oil palm expansion, in Indonesia, have been thoroughly documented in the academic literature (Frankenberg et al. 2005).

10.2.4.5 Land Tenure, Displacement, and Social Conflicts

Land tenure conflicts related to oil palm expansion have been a much-debated topic (e.g., Nesadurai 2013; Dhiaulhaq et al. 2014). In Indonesia, land conflict related to oil palm plantations is a much-debated topic, with some authors reporting the concentration of land to powerful actors and the loss of land rights through coercion/

lack of information (Cotula et al. 2008). There have also been allegations that logging companies and large plantations owners have displaced indigenous people when establishing new oil palm plantations (USAID 2009; Winrock 2009). In some cases oil palm plantations have been established without recognition of traditional land borders, rights, and interests (WWF 2006; Tata et al. 2010). Feintrenie et al. (2010) have also documented instances of social conflicts emerging in oil palm landscapes between local communities, transmigrants, and oil palm companies. Finally, the oil palm boom may have spurred speculative land acquisitions or even land grabs (McCarthy et al. 2012b) with the transfer of tenure from local communities to large companies occasionally affecting the land tenure rights of women (White and White 2012; Oosterveer et al. 2014).

10.3 Methodology

10.3.1 Methodological Approach

Section 10.2 has shown how palm oil biodiesel production/use can affect the ecosystem services upon which local populations depend for their well-being. Whether these impacts are positive or negative[7] does not only depend on the environmental and socioeconomic context of production/use but also on the technological processes and the policy instruments adopted during biodiesel production and trade (Gasparatos et al. 2011). As a result it is often very difficult to provide clear-cut answers regarding palm oil's impact on ecosystem services and human well-being and as an extension on palm oil biodiesel's sustainability.

There is a consensus that when assessing the impact of human activity on ecosystem services, it is important to integrate different types of knowledge in order to increase the effectiveness and transparency of decision-making (MA 2005b; TEEB 2010; Díaz et al. 2015). In fact the values and priorities of relevant stakeholder groups need to be captured and understood in order to better inform ecosystem services management. As a result, a wide range of participatory mechanisms has been developed both to enhance stakeholder participation and facilitate social learning at different stages of the decision-making process (e.g., Reed 2008; Stringer et al. 2006). Several studies have mapped the sociopsychological dynamics for selected communities. For example, Raymond et al. (2009) have proposed a "community values mapping method" for identifying community values and threats to natural capital assets and ecosystem services. Cuppen et al. (2010) have explored how stakeholder dialogue about different energy options in the Netherlands can bring diverse perspectives into the debate as a necessary precondition for understanding the complex bioenergy issues within the country.

[7] On several occasions the findings regarding palm oil biodiesel impacts are contradictory, for instance, the case of biofuel-related GHG emissions (refer Section 10.2.2.2).

Sections 10.4 and 10.5 capture and analyze the perceptions of various stakeholders associated with the palm oil industry about the impacts of the sector on ecosystem services, biodiversity, and human well-being. The studied impacts reflect the main impacts outlined in Sect. 10.2. The two main data collection mechanisms are:

1. Semi-structured interviews with RSPO-certified firms in Malaysia and Indonesia
2. Structured questionnaire administered to attendants of the 8th Roundtable Meeting of the RSPO (RSPO-RT8) in Jakarta in November 2010

We focused on informants affiliated in different capacities within the RSPO process, as this is a multi-stakeholder alliance that aims to promote the sustainability of palm oil. It is possibly the only large-scale and comprehensive effort to bring together the different stakeholders within the palm oil supply chain. In this respect the RSPO process has the capacity to influence the global debate on palm oil sustainability. To our best knowledge, there are no studies that empirically capture the perceptions of different RSPO stakeholders as they relate to the impacts of palm oil on ecosystem services and human well-being. Capturing and understanding the multiplicity of these perceptions is particularly relevant for the RSPO, as these differences can complicate the attainment of consensus (Schouten and Glasbergen 2011; Partzsch 2009; Silva-Castaneda 2012; von Geibler 2013).

10.3.2 Data Collection

10.3.2.1 Expert Interviews with RSPO-Certified Firms

We undertook semi-structured interviews with managers and executives involved in different capacities in the palm oil sector (e.g., oil palm plantations, trading companies, CPO plants, agricultural production in-house researchers, and refiners). These interviews were conducted during several field visits in Sumatra (2–5 July 2010), Singapore, and Johor (Malaysia) (30/June/2010 and 05/July/2010). In addition, we carried out informal discussions with workers and local residents during the field visits in these areas. Interviewees were asked questions regarding the perceived changes due to the adoption of certification standards by the palm oil industry and their impacts on ecosystem services, human well-being, and biodiversity.

10.3.2.2 RSPO-RT8 Stakeholder Survey

The structured questionnaire consisted of 15 questions, 14 of which were fixed-range questions that aimed to capture the stakeholders' perception about the impact of the oil palm industry on ecosystem services, biodiversity, and human well-being. The remaining question aimed to identify whether the respondents were involved/interested in palm oil biodiesel production.

The 14 fixed-range questions represented the key impacts of palm oil production on ecosystem services, biodiversity, and human well-being as identified in the literature review (Sect. 10.2). These questions can be divided along three impact categories:

Ecosystem services:	Freshwater services (provisioning and regulating services), air quality regulation (regulating services), erosion regulation (regulating service), soil fertility (supporting service), and climate regulation (regulating service)
Human well-being:	Income, health, land titles, labor standards, social conflicts
Biodiversity:	One dedicated question about biodiversity loss and three questions regarding some of its main drivers (i.e., invasive species, waste/pollution, and agrochemical use)

These impacts overlapped very well with the main criteria/principles of the RSPO certification scheme (RSPO 2007). For each of these questions the respondents had to provide a score between "0" and "3" that represented the difficulty of addressing the respective impact. Each of these scores was accompanied by the following short explanations:

- *0 = not difficult or costly*
- *1 = problematic but success has been achieved*
- *2 = solutions seem to exist, yet barriers prevent success*
- *3 = deep concern about finding feasible solutions*

The questionnaire also captured the stakeholders' perceptions about the main reasons behind the current impacts of palm oil production. It identified the main barriers, opportunities, and appropriate ways to reduce the negative impacts. The respondents elaborated their quantitative answers in designated boxes next to each of the 14 impact questions. The wording used as the caption in these boxes was: "Why? Please explain briefly (for example, high/low cost, political constraints/opportunities, technical feasibility/difficulty, etc).

Five hundred hardcopies of the questionnaire (450 in English and 50 translated in Bahasa Indonesia) were distributed during the plenary sessions of RSPO-RT8 (Jakarta, Indonesia), which were sufficient to cover all potential attendees during the first day plenary.[8] A brief announcement was made by the RSPO-RT8 organization committee to inform the participants about the survey and to request their collaboration in filling it.

[8] According to the official registration list, 650 people registered for the event.

10.4 Results

10.4.1 Interviews with RSPO-Certified Firms

According to the interviewees, the main management changes that companies were obliged to undertake in order to comply with the RSPO standards are to:

- Build the certification standard upon their existing ISO 14001 standard (for those companies that were already ISO 14001 certified).
- Create a sustainability team (i.e., sustainability managers, officers, directors).
- Introduce good environmental practices (such as recycling and use of byproducts), which saved costs in some cases.
- Collect data and document properly their environmental, labor, and social practices.
- Improve labor standards (e.g., safety[9]) and the social aspects of their business practices (e.g., informality in the labor force).
- Include biodiversity concerns in the management of their plantations.

The ease of compliance with the RSPO standards varied between companies and depended to a great extent on the production practices they had already adopted. Companies that already employed good management practices required relatively few changes to comply with the RSPO requirements. For some aspects, such as pesticide use or work safety standards, complying with RSPO standards did not require any modification in the existing management practices of the plantations, as the company was already "doing things beyond the requirements."

For example, during the interviews, a large producer mentioned that it is their standard business practice to consult their clients once a year about the modifications that need to be done in their current operations. As early as the early 2000s, their clients indicated a preference for adopting the ISO 14001 standard. Subsequently the company took part in the early RSPO discussions (2004) and was one of the first to obtain RSPO certification (2008) mainly due to concerns over deforestation that were articulated by their European customers. By that time the company was already ISO14001 certified and had adopted good social practices. As a result it was relatively straightforward for them to comply with the RSPO standard. It was also expressed that the RSPO certification process can be achieved more easily for large companies that are involved in several stages of the palm oil value chain (i.e., production, shipping, refining).

Some of the good environmental practices adopted by the producers during the RSPO certification process can have a ripple effect on ecosystem services and biodiversity conservation. One such example is the conservation of riparian forests. According to Malaysian legislation, riparian forests should be protected accordingly, but that is rarely the case. However, the RSPO standards require the conservation of riparian forests, which are essentially buffer zones between streams and

[9] RSPO-certified companies are now providing training and safety equipment to their employees.

human dominated landscapes. Such forests can host several animal species and provide ecosystem services such as water purification and flood control (MA 2005a). A second good environmental practice promoted by the RSPO is the prohibition of planting oil palm trees in steep terrains (25 degrees or more). Forests in steep terrains could provide a series of ecosystem services including erosion control (MA 2005a). Finally companies that aspire for RSPO certification are required to use only certain low-toxicity RSPO-approved agrochemicals. At the same time, the RSPO is quite friendly toward the incorporation of more sustainable forms of pest management.[10] Shift toward less hazardous forms of pest management can offer several benefits given the generally negative impact of agrochemicals on ecosystem services (e.g., on freshwater services), human well-being (e.g., health), and biodiversity (refer to Sect. 10.2).

It should be mentioned that there are several occasions where the RSPO certification process has encountered obstacles by palm oil-producing companies. Our interviews suggested that it is not uncommon for site managers to be apprehensive of the RSPO certification scheme fearing potential productivity decreases. Furthermore companies need to invest resources in order to comply with the RSPO standard, which increases the cost of palm oil production (refer to Sect. 10.5).

Interviewees also noted that the environmental (and indirectly the social) concerns addressed by the RSPO system or any similar certification scheme is not the only or most efficient way to promote the sustainability of the oil palm sector. Researchers from the development center of an interviewed company pointed out that new technologies in plant breeding are the "most promising instrument to make sustainable palm oil a reality." According to the interviewee, the current productivity of palm oil plants has not reached the maximum potential, and hybridization technology will most probably bring astonishing increases in yields (up to sevenfold) in the coming years. This will eliminate the need for expansion into ecosystems rich in ecosystem services and biodiversity and in turn will allow smallholders to make a living within current land cultivation areas.

In addition, top managers noted that, while RSPO and other similar certification schemes might make a big difference for those companies currently having a low environmental and social performance, some companies are moving "beyond certification." That is, in order to gain a competitive advantage by becoming sustainability leaders within the sector, certain companies are investing heavily in making their operations "truly sustainable." According to the interviewee, environmental and social concerns regarding palm oil "are here to stay"; therefore "sooner rather than later" firms who do not take steps toward improved performance will "lose out."

---

[10] For example, companies plant flowers in order to attract insects (e.g., bees) that hatch their eggs in the larvae of the worm that destroys the oil palm leaves or use owls to hunt rodents that climb on palm trees to eat the palm kernels.

Table 10.2 Characteristics of respondents

Grouping	Variable	Responses	Fraction (%)
Stakeholder group	Smallholders	13	10.2
	Large producers	56	44.1
	Consultancy	13	10.2
	Processor	6	4.7
	Distributor	2	1.6
	Certifier	7	5.5
	Final consumer	7	5.5
	Investor	3	2.4
	NGO	8	6.3
	Government	0	0.0
	Academia	5	3.9
	Other/NA	7	5.5
	Total	**127**	**100.0**
Region	Indonesia	53	41.7
	Malaysia	25	19.7
	Singapore	3	2.4
	Thailand	3	2.4
	Other/NA	43	33.9
	Total	**127**	**100.0**
RSPO-certified	Yes	36	28.3
	No	43	33.9
	NA	48	37.8
	Total	**127**	**100.0**

10.4.2 RSPO-RT8 Stakeholder Survey

A total of 139 questionnaires were returned with the response rate estimated at 25.3% based on the average number of attendees per plenary session (550 persons according to in situ observations). The response rates for each individual question varied across the 14 fixed-range questions, with questions in the front page of the questionnaire being answered more frequently than those on the back side. Quantitative responses to all 14 questions were included in 86 questionnaires, while the response rate to the question regarding the involvement/interest of the respondent in palm oil biodiesel was markedly lower.

10.4.2.1 Quantitative Analysis

In total 127 questionnaires were used for the quantitative analysis. Most of these respondents were representing actual oil palm producers (Table 10.2). Respondents representing "large producers" and "small producers" accounted for 54.3% of our sample with the remaining representing different organizations further down the

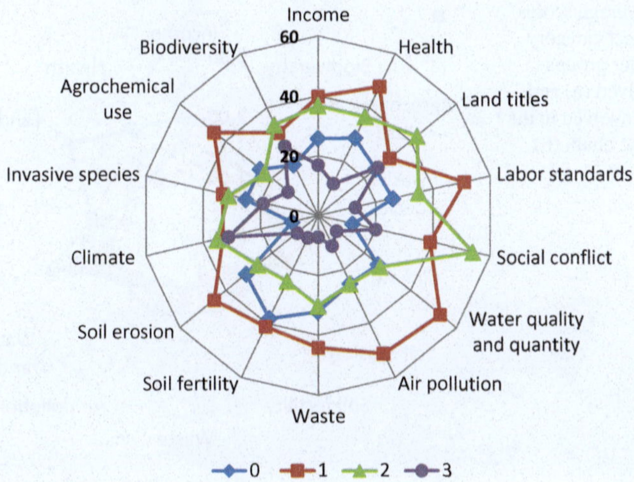

Fig. 10.4 Distribution of scores among the 14 impact categories

production/consumption chain (11.8%). It must be noted that several individuals represented organizations that are not directly involved in the palm oil industry but had a vested interest in the palm oil sector, e.g., "NGOs," "academics," "consultancies," and "investors." Most of the respondents were affiliated with entities that were based or were operating in Indonesia, Malaysia, Singapore, and Thailand. Several more respondents identified their field of operation in the broad region with very few respondents explicitly stating an affiliation outside Southeast Asia. This is a good indicator that our sample had some understanding of the specific environmental and socioeconomic conditions surrounding oil palm agriculture in the region. Finally, 28.3% of the respondents represented organizations that were RSPO-certified with 44% of the rest intending RSPO certification.

For all questions, the most frequent scores reported were "1" and "2," with "0" and "3" being reported less frequently (see Fig. 10.4). This suggests that there is a widespread perception among the respondents that the palm oil industry has indeed a negative impact on ecosystem services, biodiversity, and human well-being but that solutions exist and that progress has been (or can be) achieved. This is not surprising considering that most of the respondents were affiliated with organizations that were either RSPO-certified or were intending to become RSPO-certified in the future. As explained in the previous section, the compliance with RSPO certification standards implies the adoption of certain good management practices that have a ripple effect on ecosystems and human well-being.

It is worth mentioning that for only four issues (i.e., "land titles," "social conflict," "climate regulation," and biodiversity), the most popular score was "2." Most importantly "climate regulation" and "biodiversity" were the only issues for which there was an almost equal response rate for scores "1," "2," and "3." Such polarized views suggest the controversial nature of both topics with a lack of consensus within the RSPO community.

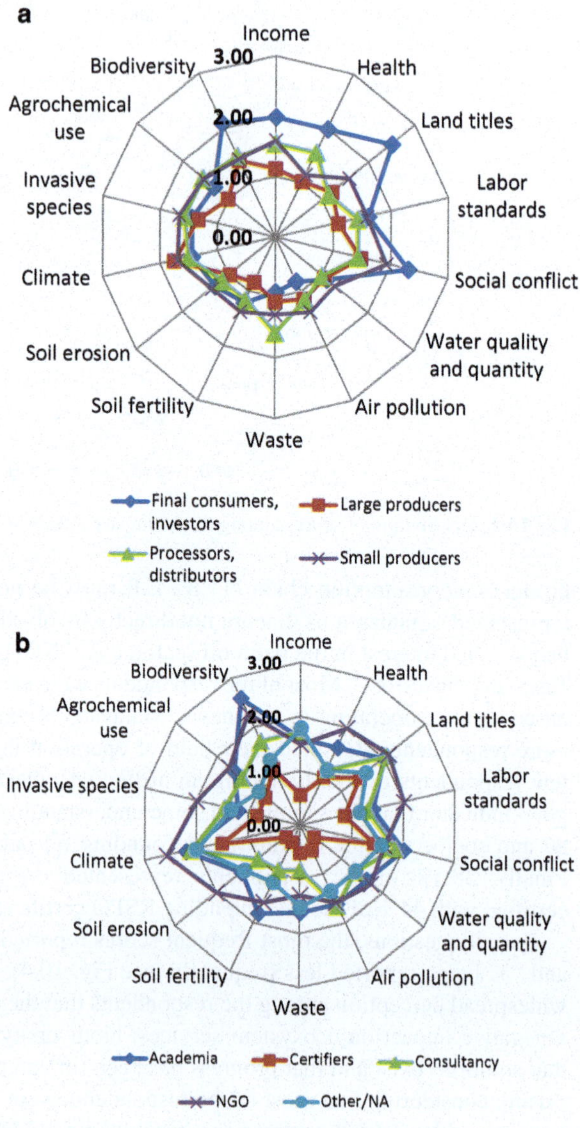

Fig. 10.5 Average scores for each impact category for stakeholder groups directly involved (**a**) and not directly involved in the palm oil value chain (**b**)

On the other hand, "0" was the second most popular response for only three impact categories: "soil fertility," "soil erosion," and "agrochemical use." This trend might have been due to the realization by oil palm producers of the importance of healthy soil for the sustainability of their firms and as a result taking the necessary actions to maintain good soil quality prior to the launching of the RSPO certification scheme.

Figure 10.5a, b highlights the average scores for each of the 14 impact categories by stakeholder group. The different stakeholder groups can be roughly divided into two categories: those groups that are directly involved in the palm oil production/consumption value chain (Fig. 10.5a) and those that are not directly involved but have a vested interest in the sector (Fig. 10.5b). Generally speaking the stakeholders that are directly involved in the palm oil value chain have provided lower scores (on average), which seems to imply optimism about the potential to tackle the negative impacts of palm oil production on ecosystem services, biodiversity, and human well-being.

However there are two very interesting exceptions. Certifiers are not directly involved in the palm oil value chain but have reported the most optimistic responses. In particular they provided the lowest average scores in 12 of the 14 impact categories and the second lowest in 1 impact category. This optimism might stem from the certifiers' specific role in the sector, which is essentially to assess whether palm oil producers meet the RSPO standards and as such whether they merit to be awarded an RSPO certification. As it has been discussed in Sect. 10.4.1, meeting the RSPO criteria is not necessarily difficult and should be seen as a first step toward sustainable palm oil production. It could be argued that the optimistic perception of certifiers reflects their role in the RSPO certification process which is precisely to make it accountable and thus have faith that specific interventions can indeed reduce the negative impacts of palm oil production. The second most optimistic group was the "large producers," which reported the second lowest score in 9 of the 14 impact categories.

Another interesting exception is "final consumers."[11] Even though "final consumers" are directly integrated in the palm oil value chain, they have provided the highest average scores in four of the five human well-being impacts and the second highest for the remaining impact. More interestingly their scores for ecosystem services and biodiversity impacts are much lower and more in line with the responses provided by the other stakeholder groups directly involved in the palm oil value chain. This trend might reflect the escalating social concerns among consumers (especially in countries of North America and Europe) regarding the social practices of commodities production in the global South. "NGOs" are also highly skeptical of the progress that can be achieved by the sector. They provided the highest score in seven impact categories and the second highest in three. "Academics" are also highly skeptical, albeit more moderate than "NGOs," providing the highest average scores in three impact categories and the second highest score in seven.

Among stakeholder responses that are statistically significant (at the 10th percentile), "NGOs" consistently perceive greater challenges to achieve sustainability than "large producers" do. This confirms to some extent the stereotype that "NGOs"

[11] This category includes companies that use palm oil as a raw material and are active in diverse industries such as food processing, pharmaceuticals, and cosmetics. "Final consumer" concerns over palm oil sustainability have been identified as a major driving force behind the development and increasing adoption of the RSPO certification scheme.

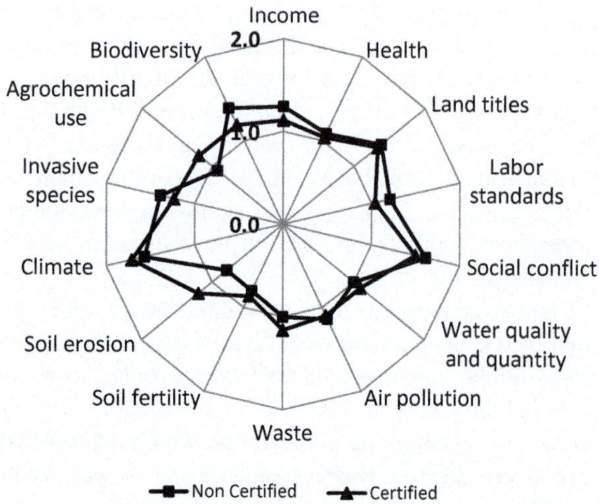

Fig. 10.6 Average scores for producers ("large producers" and "small producers") according to their RSPO certification status

and "large producers" represent the two extremes viewpoints in the debate about palm oil sustainability.

Figure 10.6 shows that the perception of palm oil producers about the impacts of their business is the same whether they have obtained RSPO certification or not. This finding is quite interesting and implies that oil palm producers do not change their perceptions significantly during their involvement in the RSPO process or the roundtable discussions that are conceived as a way to promote the dialogue of stakeholders with different perspectives. This might imply that oil palm producers are actually seeking RSPO certification for pragmatic reasons such as the improvement of their image. Such a conclusion would be consistent with some of the findings that were obtained during the interviews with RSPO-certified producers (refer to Sects. 10.4.1 and 10.5).

In a similar manner the perceptions of "large producers" and "small producers" based in Indonesia and Malaysia are relatively similar (Fig. 10.7). This is somewhat surprising, given the reputation that the Malaysian palm oil sector is subject to stricter regulation. The main difference lies on the perceptions about land titles. Respondents from Indonesia provide a higher average score, which reflects a concern that is particularly pertinent in the country. Sometimes the customary rights to land use are different than the legal rights, and this has led to conflicts between producers and local communities (Sect. 10.2.4.5).

Finally the perception of respondents with an interest in biodiesel production exhibits a very similar pattern when compared to the perceptions of respondents that are not interested (Fig. 10.8). This might be due to the fact that the final use of palm oil is determined further up in the supply chain. Besides, oil palm for biodiesel was still at the time of the survey a relatively incipient business; thus strong opinions might not have been formed yet with respect to this issue.

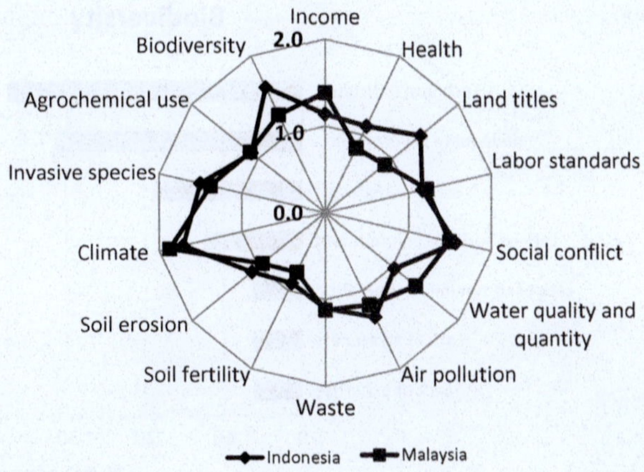

Fig. 10.7 Average scores for producers ("large producers" and "small producers") that are based in Indonesia and Malaysia

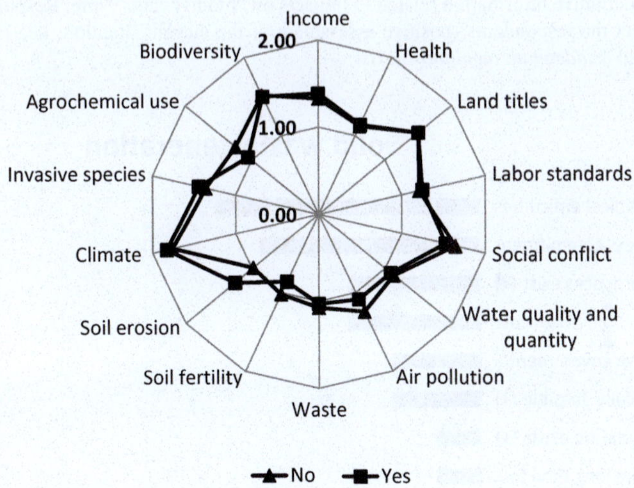

Fig. 10.8 Average scores for stakeholder "interested" and "not interested" in palm oil biodiesel production

10.4.2.2 Qualitative Analysis

Response rates were much lower for the qualitative part of the questionnaire and varied significantly among the different impact categories. As expected the obtained results were quite different for each impact (Figs. 10.9, 10.10, 10.11, 10.12, 10.13, 10.14, 10.15, 10.16, 10.17, 10.18, 10.19, 10.20, 10.21, and 10.22), but some common trends can be identified.

Surprisingly, with the exception of "health," cost has not been identified as a major barrier for improving the impacts of palm oil on human well-being (Figs. 10.18, 10.19, 10.20, 10.21, and 10.22). On the other hand, cost and technical

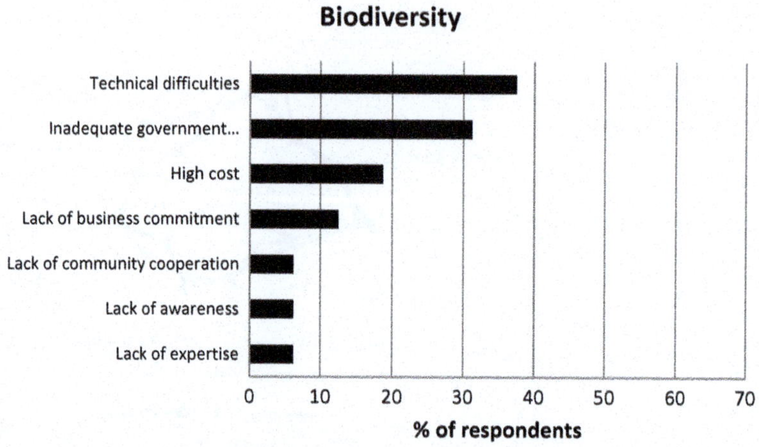

Fig. 10.9 Qualitative information related to impacts on "biodiversity." Note: Responses that contain (+) reflect the respondents' positive assessment of the current situation, i.e., *(+) regulation* should be read as *adequate regulations exist*

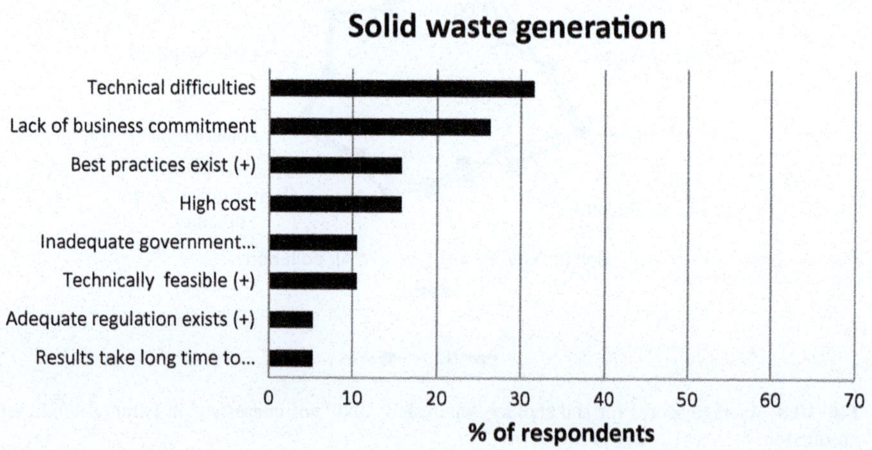

Fig. 10.10 Qualitative information related to impacts of "solid waste generation." Note: Responses that contain (+) reflect the respondents' positive assessment of the current situation, i.e., *(+) regulation* should be read as *adequate regulations exist*

constraints were identified as the key barriers for reducing the stress on ecosystem services and biodiversity (Figs. 10.9, 10.10, 10.11, 10.12, 10.13, 10.14, 10.15, 10.16, and 10.17). That is to be expected given that most of these impacts can be ameliorated with the adoption of good management practices (refer to Sect. 10.2.2) which are usually more costly than the current practices of the palm oil industry.

Another interesting finding is that for each impact category, government or government-related issues (e.g., regulations, implementation/enforcement of

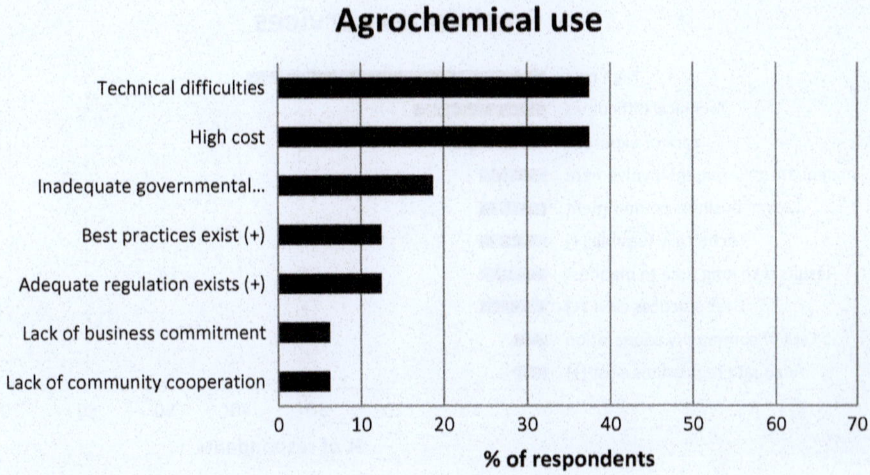

Fig. 10.11 Qualitative information related to impacts of "agrochemical use." Note: Responses that contain (+) reflect the respondents' positive assessment of the current situation, i.e., *(+) regulation* should be read as *adequate regulations exist*

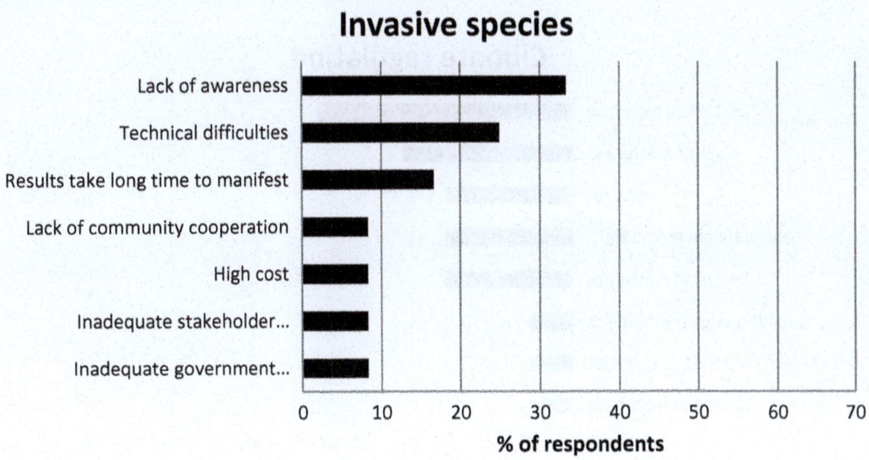

Fig. 10.12 Qualitative information related to impacts of "invasive species"

regulations) were identified as a major contributor to the current situation and a key barrier for future improvement (Figs. 10.9, 10.10, 10.11, 10.12, 10.13, 10.14, 10.15, 10.16, 10.17, 10.18, 10.19, 10.20, 10.21, and 10.22). At the same time none of our respondents was affiliated with a governmental body. In fact very few RSPO-RT8 participants represent government institutions, suggesting the lack of proper governmental engagement in the RSPO-RT8 process.

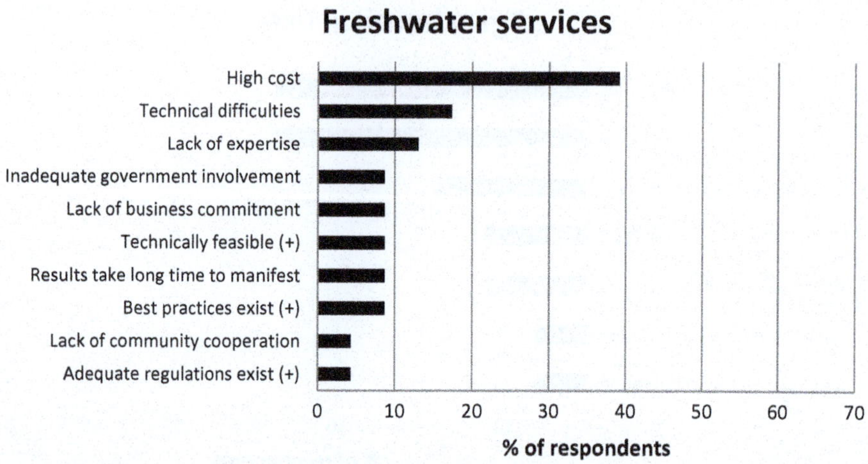

Fig. 10.13 Qualitative information related to impacts on "freshwater services." Note: Responses that contain (+) reflect the respondents' positive assessment of the current situation, i.e., *(+) regulation* should be read as *adequate regulations exist*

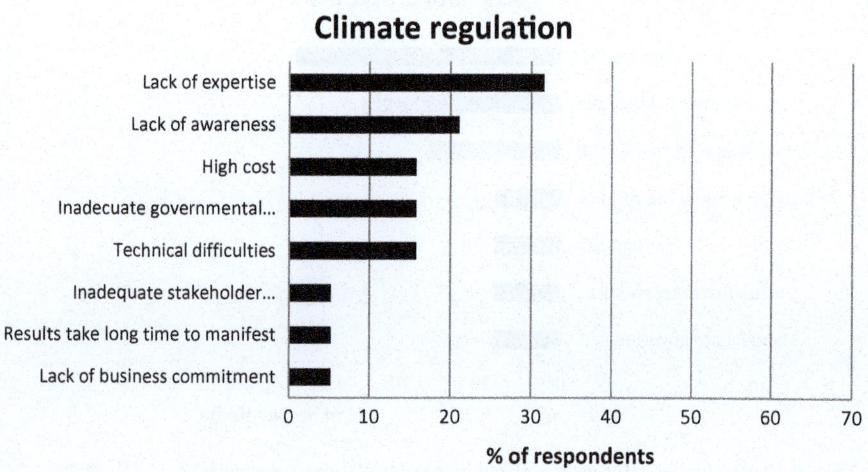

Fig. 10.14 Qualitative information related to impacts on "climate regulation." Note: Responses that contain (+) reflect the respondents' positive assessment of the current situation, i.e., *(+) regulation* should be read as *adequate regulations exist*

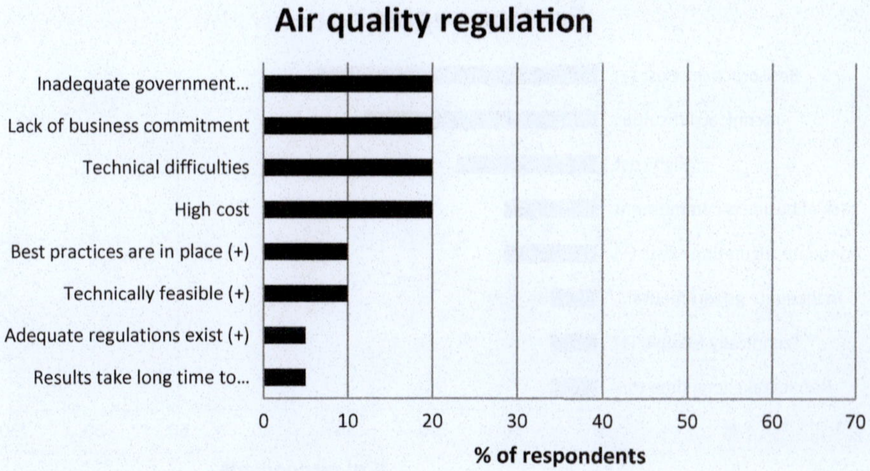

Fig. 10.15 Qualitative information related to impacts on "air quality regulation." Note: Responses that contain (+) reflect the respondents' positive assessment of the current situation, i.e., *(+) regulation* should be read as *adequate regulations exist*

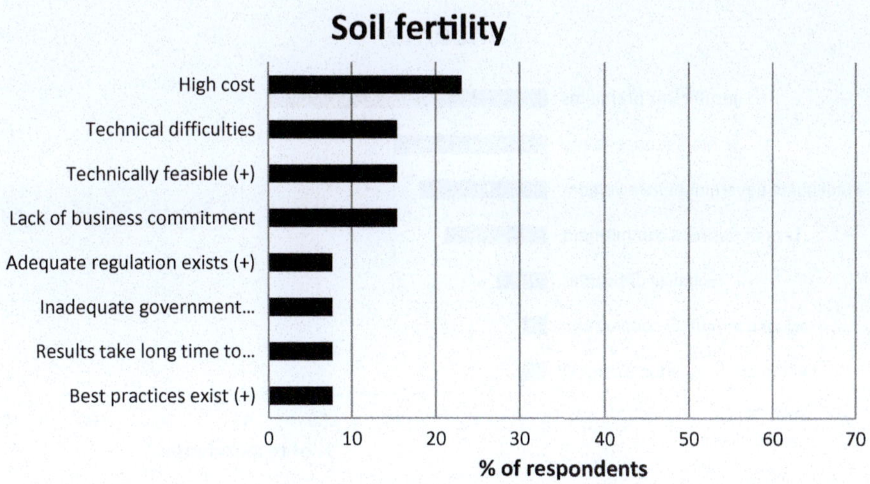

Fig. 10.16 Qualitative information related to impacts on "soil fertility." Note: Responses that contain (+) reflect the respondents' positive assessment of the current situation, i.e., *(+) regulation* should be read as *adequate regulations exist*

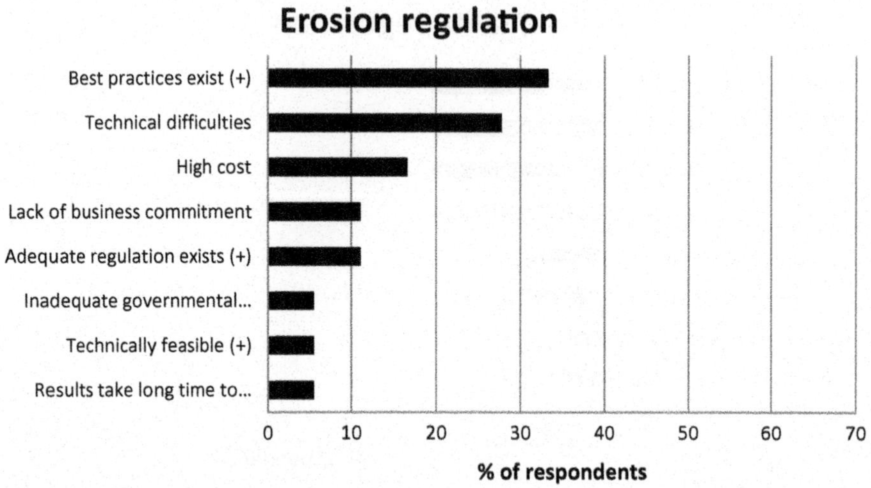

Fig. 10.17 Qualitative information related to impacts on "erosion regulation." Note: Responses that contain (+) reflect the respondents' positive assessment of the current situation, i.e., *(+) regulation* should be read as *adequate regulations exist*

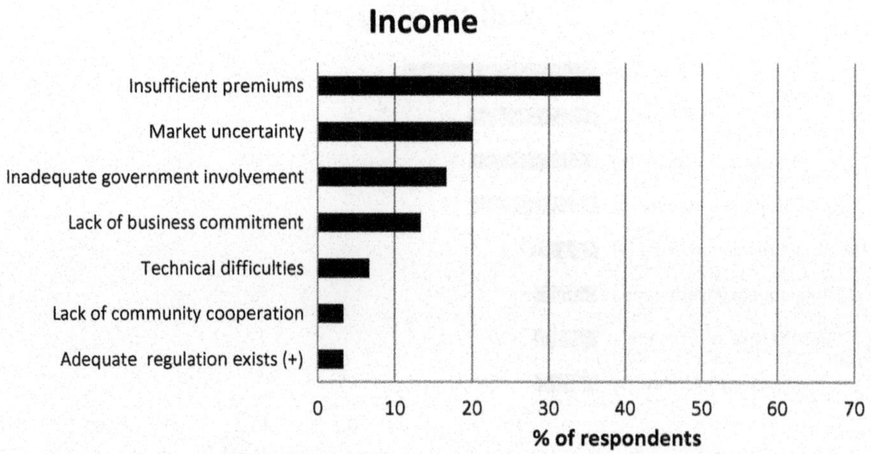

Fig. 10.18 Qualitative information related to impacts on "income." Note: Responses that contain (+) reflect the respondents' positive assessment of the current situation, i.e., *(+) regulation* should be read as *adequate regulations exist*

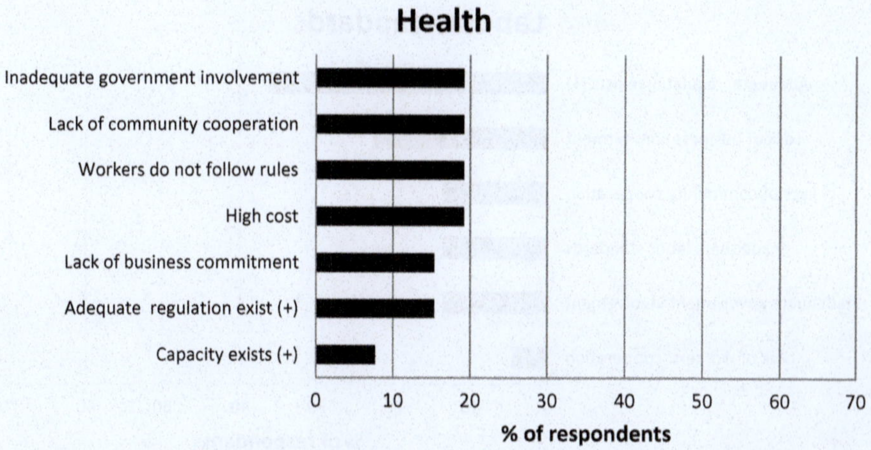

Fig. 10.19 Qualitative information related to impacts on "health." Note: Responses that contain (+) reflect the respondents' positive assessment of the current situation, i.e., *(+) regulation* should be read as *adequate regulations exist*

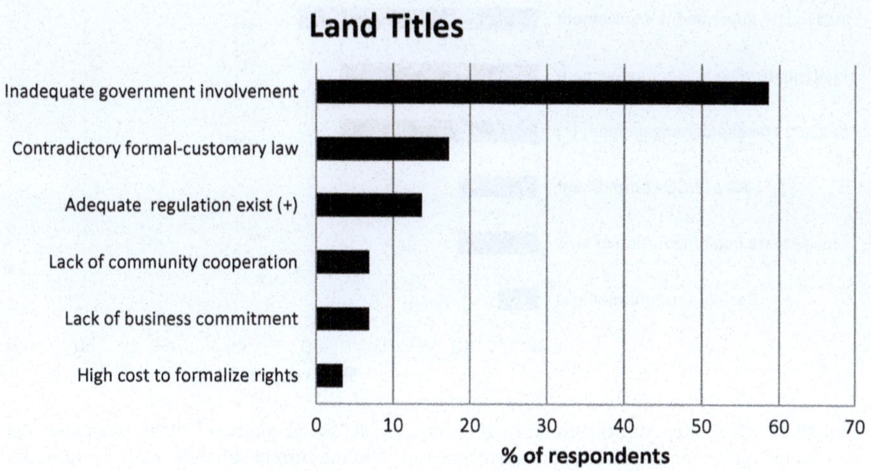

Fig. 10.20 Qualitative information related to impacts on "land titles." Note: Responses that contain (+) reflect the respondents' positive assessment of the current situation, i.e., *(+) regulation* should be read as *adequate regulations exist*

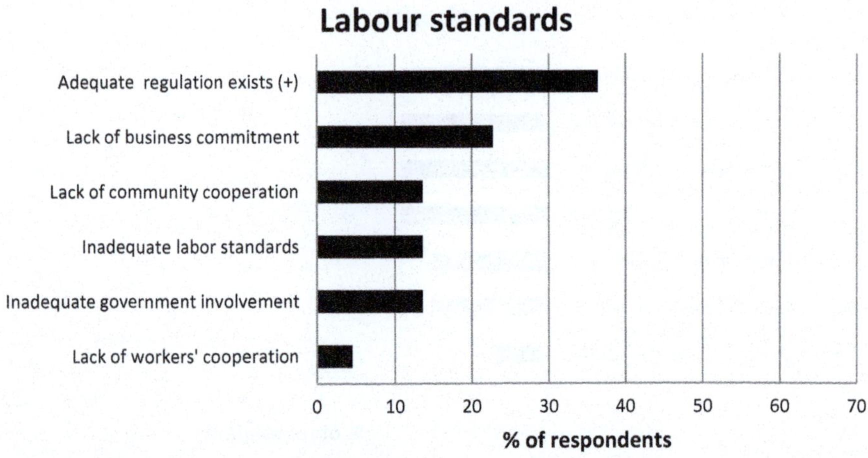

Fig. 10.21 Qualitative information related to impacts on "labor standards." Note: Responses that contain (+) reflect the respondents' positive assessment of the current situation, i.e., *(+) regulation* should be read as *adequate regulations exist*

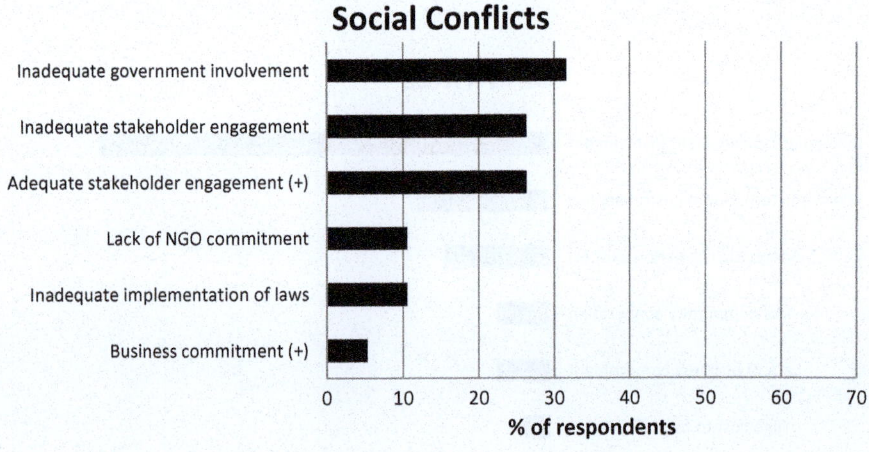

Fig. 10.22 Qualitative information related to impacts on "social conflicts." Note: Responses that contain (+) reflect the respondents' positive assessment of the current situation, i.e., *(+) regulation* should be read as *adequate regulations exist*

10.5 Discussion

In 2010, approximately 4.5% of the global palm oil production capacity was RSPO-certified, while in 2012 approximately this stood at 11% (RSPO 2012). Within the following 5 years, final consumers from Europe had pledged to purchase at least 40% of their palm oil from RSPO-certified companies. One of the driving forces behind this increasing interest in RSPO-certified palm oil is the current debate on biofuel-induced deforestation in Southeast Asia and its accompanying impacts (i.e., biodiversity loss, GHG emissions). The CEO of a large palm oil production company referred to this social pressure by implying that "environmentalists are so obsessed with palm oil production that they forgot the illegal timber industry, which still is the main driver of deforestation".

The adoption of the RSPO standards can produce a series of environmental and social benefits both locally and globally. Compliance with the RSPO standard implies the adoption of certain socially and ecologically responsible practices. Some of these practices can promote environmental quality and subsequently enhance the provision of certain ecosystem services. Additionally RSPO-certified firms are required to monitor and document regularly their performance. Such activities usually entail higher production costs. In fact, managing a RSPO-certified plantation can cost RMM 20/ha/yearr. (USD 6.3/ha/year) more than a conventional farm. Despite this interest the market does not recognize RSPO-certified palm oil with a good premium (just USD 5–8/ton, with CPO prices hovering around USD 800/ton).

At the time of the survey, palm oil biodiesel production was a new business option, and shifting to it was not always a feasible investment strategy. For example, a large palm oil producer in Bohor (Malaysia), with decades of experience in the industry, had laid careful plans to commence biodiesel production. However when the production was about to start, the company encountered problems that mainly stemmed from European concerns over potential "food vs. fuel" conflicts, as well as the decreasing petroleum prices and the increasing CPO prices (July 2010). In a way they were "forced" to halt their biodiesel investment plans fearing that their venture into biofuels might hurt their well-establish business in the food sector. Even though they were considering selling the biofuel conversion plant, they were also considering obtaining RSPO certification for some of their plantations in order to guarantee the good image of the company. In fact the adoption of RSPO certification is seen by several companies as a means of improving their image and as a prerequisite to secure a competitive advantage in the EU, the USA, and Japanese markets, whether for food or biodiesel production.

With these things in mind, the main question that remains is: will RSPO certification, in its current format, be enough to guarantee palm oil sustainability in the long run? In our view there are three main reasons why this will become more difficult in the future.

First of all, a large portion of the Malaysian/Indonesian palm oil exports goes (and will go) to China and India, two countries that have a lax approach toward environmental standards. Secondly, as biodiversity loss and climate change have become the most dominant environmental concerns in the palm oil sustainability discourse, the RSPO will be under increasing pressure to design more effective strategies to tackle these issues. However as the results of our survey have shown, biodiversity loss and climate change are the two impacts with the most polarized perceptions regarding the extent to which the sector can tackle them in a cost-effective manner. If the RSPO adopts standards that are not cost-effective, then palm oil producers might avoid obtaining RSPO certification. Thirdly, and maybe most importantly, as a management system, the RSPO can guarantee minimum social and environmental standards of palm oil production but cannot put a cap on palm oil production capacity. All current scenarios suggest that the demand for palm oil will grow in the future and there is little doubt that this added demand will be an agent of deforestation in the region.

On the other hand, it should be kept in mind that oil palm agriculture is a highly efficient form of vegetable oil production. Palm oil yields can range from 4–5 ton/ha to 9 ton/ha when using good management practices in favorable growing conditions. Other oil crops have much lower yields (FAO 2017) which means that their expansion will most likely be a major factor of LUCC elsewhere, possibly degrading ecosystem services and biodiversity.

As already discussed the RSPO certification scheme is an important governance mechanism that through the improvement of local ecological and social conditions can have ripple positive effects on local ecosystem services and human well-being. However, a long-term perspective on the sustainability of the sector needs a regional or global governance mechanism to address all the possible trade-offs particularly when considering the increasing demand for palm oil biodiesel and the emergence of climate change and biodiversity loss as overarching policy agendas.

Some suggestions that can further improve the sustainability of palm oil biodiesel production in Southeast Asia are:

- Starting from a reference year, create a cap on oil palm plantations on forested land but allowing the establishment of plantations in areas that were formerly under other uses (e.g., agriculture, pasture, degraded land).
- Improve the efficiency of palm oil production, perhaps by developing new high-yield oil palm varieties.
- Promote and improve the efficiency of small-scale biofuel projects (e.g., FAO 2009).
- Promote agroforestry and small-scale palm oil production.
- Create a "certificate of origin" for the biodiesel originating from plantations located in administrative regions (municipalities) with controlled deforestation. For example, a farm would get the certificate only if the municipality where it is located has a small or zero rate of deforestation. This would put peer pressure over the producers in the municipality.

Finally, it should be mentioned that using stakeholders' perceptions as a means of understanding the impacts of oil palm expansion on ecosystem services, human well-being, and biodiversity implies the reliance on subjective and thus unverifiable data to infer the actual impacts and their magnitude. On the other hand, it allows for a relatively quick identification and evaluation of potential impacts which would have been too time/resource consuming to perform otherwise. Therefore, the results of our study should not be viewed as an attempt to replace empirical ecosystem impact assessment. Our results simply aim to understand stakeholders' values and priorities in order to (a) identify and (b) quickly assess the main impacts of oil palm expansion on ecosystem services, human well-being, and biodiversity. Such information can complement empirical evidence and can be used to inform the development of appropriate policy mechanisms that will target palm oil sustainability.

10.6 Conclusions

Our study has reviewed the evidence of how the production and use of palm oil biofuels affects biodiversity and several provisioning (e.g., fuel, food, freshwater), regulating (e.g., climate regulation, air quality regulation, erosion control), and cultural ecosystem services in Malaysia and Indonesia. It also discussed how this change in ecosystem services provision can affect the human well-being of the local communities.

The surge in biofuels in the late 2000s put some pressure on the palm oil industry to improve the sustainability performance of its product. For example, the European Union through its 2009/28/EC Directive has adopted legally binding provisions about the acceptable limits of GHG emissions and deforestation for palm oil that is used for energy purposes within Europe. Such provisions have largely catalyzed a growing demand for the adoption of certification standards that was a promising sign about the sector's impact on ecosystem services.

On several occasions voluntary market-driven mechanisms such as the RSPO certification scheme have successfully promoted good environmental and social practices. The demand to conserve riparian forests, prohibit the cultivation of oil palms on steep terrains, and phase out toxic agrochemicals are only some of the good practices "required" by the RSPO standards. These practices can have a ripple positive effect on ecosystem services and biodiversity conservation.

At the same time, certification schemes face great challenges. Using the RSPO as an example, this chapter has highlighted that this scheme might not be sufficient to tackle palm oil's intertwined impacts on climate change, ecosystem services, and biodiversity. There is the need to have a dynamic approach to certification schemes, adjusting them to account for the aggregate stress of the growing oil palm sector. A balance has to be struck between acceptable environmental targets and cost-effectiveness in order to ensure the acceptability of certification standards by producers.

However, far from being homogeneous, the perceptions about the sector's impacts on ecosystem services, biodiversity, and human well-being, (and the potential to mitigate them), were very divergent in the early 2010s among the different actors that comprise this multi-stakeholder alliance. Furthermore, stakeholders' perception regarding the industry's impacts on climate and biodiversity was very polarized. Reconciling these different perceptions should be a first step for coming up with mutually acceptable standards. This would be a great bet for certification schemes such as the RSPO in the future. This will most likely involve a closer cooperation between the different RSPO working groups and indeed between the diverse stakeholders themselves. It will also demand a closer cooperation between the RSPO and government institutions in Malaysia and Indonesia.

Acknowledgments This research was part of the project "Biofuel Use Strategies for Sustainable Development" supported by the Global Environment Research Fund (E-0802) of the Ministry of the Environment, Japan. We also thank the United Nations University Institute of Advanced Studies (UNU-IAS) for the support for the research.

References

Actionaid (2010) Meals per gallon: the impact of industrial biofuels on people and global hunger. Actionaid, London

APEC (2010) Indonesia Biofuels Activities. Asia-Pacific Economic Cooperation. Accessed on 6 Feb 2011 at: http://www.biofuels.apec.org/me_indonesia.html#biofuels

Begum S, Kumaran P, Jayakumar M (2013) Use of oil palm waste as a renewable energy source and its impact on reduction of air pollution in context of Malaysia. IOP Conf Ser Earth Environ Sci 16:012026

Bouët A, Laborde Debucquet D (2016) Food crisis and export taxation: revisiting the adverse effects of noncooperative aspect of trade policies. In: Kalkuhl M, von Braun J, Torero M (eds) Food price volatility and its implications for food security and policy. Springer, Berlin, pp 167–179

Budidarsono S, Dewi S, Sofiyuddin M, Rahmanulloh A (2012) Socioeconomic impact assessment of palm oil production. World Agroforestry Centre (ICRAF), Bogor

Buschman FA, Hoitink AJF, de Jong SM, Hoekstra P, Hidayat H, Sassi MG (2012) Suspended sediment load in the tidal zone of an Indonesian river. Hydrol Earth Syst Sci 16:4191–4204

Cahyadi ER (2013) Contract farming and risks for smallholders in the oil-palm industry in Indonesia. Bull Indones Econ Stud 49:381–382

Cahyadi ER, Waibel H (2013) Is contract farming in the Indonesian oil palm industry propoor? J Southeast Asian Econ 30:62–76

Carlson KM, Curran LM, Ratnasari D, Pittman AM, Soares-Filho BS, Asner GP, Trigg SN, Gaveau DA, Lawrence D, Rodrigues HO (2012) Committed carbon emissions, deforestation, and community land conversion from oil palm plantation expansion in West Kalimantan, Indonesia. Proc Natl Acad Sci U S A 109:7559–7564

Cotula L, Dyer N, Vermeulen S (2008) Fuelling exclusion? The biofuels boom and poor people's access to land. International Institute for Environment and Development, London

Cramb R, McCarthy JF (eds) (2016) The oil palm complex: smallholders, agribusiness and the state in Indonesia and Malaysia. NUS Press, Singapore

Cramb RA, Sujang PS (2012) Pathways through the plantation: oil palm smallholders and livelihood strategies in Sarawak, Malaysia. Australian Agricultural and Resource Economics Society (AARES), Fremantle

Crutzen PJ, Andreae MO (1990) Biomass burning in the tropics: impact on atmospheric chemistry and biogeochemical cycles. Science 250(4988):1669–1678

Cuppen E et al (2010) Q methodology to select participants for a stakeholder dialogue on energy options from biomass in the Netherlands. Ecol Econ 69(3):579–591

Danielsen F et al (2009) Biofuel plantations on forested lands: double jeopardy for biodiversity and climate. Conserv Biol 23(2):348–358

De Fraiture C, Berndes G (2009) Biofuels and water. In: Howarth RW, Bringezu S (eds) Biofuels: environmental consequences and interactions with changing land use, proceedings of the Scientific Committee on Problems of the Environment (SCOPE) international biofuels project rapid assessment. Cornell University, Ithaca

de Vries SC, van de Ven GWJ, Ittersum MK, Giller KE (2010) Resource use efficiency and environmental performance of nine major biofuel crops, processed by firstgeneration conversion techniques. Biomass Bioenergy 34(5):588–601

Dhiaulhaq A, Gritten D, De Bruyn T, Yasmi Y, Zazali A, Silalahi M (2014) Transforming conflict in plantations through mediation: Lessons and experiences from Sumatera, Indonesia. Forest Policy Econ 41:22–30

Díaz S, Demissew S, Carabias J, Joly C, Lonsdale M, Ash N et al (2015) The IPBES conceptual framework – connecting nature and people. Curr Opin Environ Sustain 14:1–16

Dislich C, Keyel AC, Salecker J, Kisel Y, Meyer KM, Auliya M, Barnes AD, Corre MD, Darras K, Faust H, Hess B, Klasen S, Knohl A, Kreft H, Meijide A, Nurdiansyah F, Otten F, Pe'er G, Steinebach S, Tarigan S, Tölle MH, Tscharntke T, Wiegand K (2017) A review of the ecosystem functions in oil palm plantations, using forests as a reference system. Biol Rev 92:1539–1569

DTE (2005) Oil palm expansion will bring more conflicts. Down to Earth, International Campaign for Ecological Justice in Indonesia, No 66, August

EU (2009) Directive 2009/28/EC on the promotion of the use of energy from renewable sources. Off J Eur Union L140:16–62

Euler M, Krishna V, Schwarze S, Siregar H, Qaim M (2017) Oil palm adoption, household welfare, and nutrition among smallholder farmers in Indonesia. World Dev 93:219–235

FAO (2004) Fertiliser use by crop in Malaysia. Food and Agriculture Organisation, Rome

FAO (2005) Fertiliser use by crop in Indonesia. Food and Agriculture Organisation, Rome

FAO (2006) World agriculture: towards 2030/2050. Food and Agriculture Organisation, Rome

FAO (2009) Small-scale bioenergy initiatives: brief description and preliminary lessons on livelihood impacts from case studies in Asia, Latin America and Africa. Food and Agricultural Organisation, Rome

FAO (2017) FAOSTAT databases. Food and Agricultural Organisation, Rome. Accessed on 14 Nov 2017 at: http://faostat.fao.org/

Fargione J et al (2008) Land clearing and the biofuel carbon debt. Science 319(5867):1235–1238

FEDIO (2016) Vegetable oils production, imports, exports and consumption: trends 1980–2016. FEDIO, Brussels

Feintrenie L, Chong WK, Levang P (2010) Why do farmers prefer oil palm? Lessons learnt from Bungo District, Indonesia. Small-scale For 9:379–396

FIAM (2009) Evolution of fertilizer use by crop in Malaysia: recent trends and prospects. Fertilizer Industry Association of Malaysia, Shah Alam

Fitzherbert EB et al (2008) How will oil palm expansion affect biodiversity? Trends Ecol Evol 23(10):538–545

Foster WA, Snaddon JL, Turner EC, Fayle TM, Cockerill TD, Ellwood MDF, Broad GR, Chung AYC, Eggleton P, Khen CV, Yusah KM (2011) Establishing the evidence base for maintaining biodiversity and ecosystem function in the oil palm landscapes of South East Asia. Philos Trans R Soc B: Biol Sci 366:3277–3291

Fowler D, Nemitz E, Misztal P, Di Marco C, Skiba U, Ryder J, Helfter C, Cape JN, Owen S, Dorsey J, Gallagher MW, Coyle M, Phillips G, Davison B, Langford B et al (2011) Effects of land use on surface-atmosphere exchanges of trace gases and energy in Borneo: comparing fluxes over oil palm plantations and a rainforest. Philos Trans R Soc B: Biol Sci 366:3196–3209

Frankenberg E, McKee D, Duncan T (2005) Health consequences of forest fires in Indonesia. Demography 42(1):109–129

Gasparatos A, Stromberg P, Takeuchi K (2011) Biofuels, ecosystem services and human wellbeing: putting biofuels in the ecosystem services narrative. Agric Ecosyst Environ 142:111–128

Gasparatos A, Doll C, Esteban M, Ahmed A, Olang T (2017) Biodiversity and renewable energy: implications for transitioning to a green economy. Renew Sust Energ Rev 70:161–184

Gatto M, Wollni M, Qaim M (2015) Oil palm boom and land-use dynamics in Indonesia: the role of policies and socio-economic factors. Land Use Policy 46:292–303

Gerbens-Leenes PW, Hoekstra AY, van der Meer T (2009a) The water footprint of bioenergy. PNAS: Proc Natl Acad Sci 106(25):10219–10223

Gerbens-Leenes PW, Hoekstra AY, van der Meer T (2009b) The water footprint of energy from biomass: a quantitative assessment and consequences of an increasing share of bio-energy in energy supply. Ecol Econ 68(4):1052–1060

Gibbs HK, Ruesch AS, Achard F, Clayton MK, Holmgren P, Ramankutty N, Foley JA (2010) Tropical forests were the primary sources of new agricultural land in the 1980s and 1990s. Proc Natl Acad Sci U S A 107:16732–16737

Gissi E, Gaglio M, Reho M (2016) Sustainable energy potential from biomass through ecosystem services trade-off analysis: the case of the province of Rovigo (northern Italy). Ecosyst Serv 18:1–19

Guillaume T, Damris M, Kuzyakov Y (2015) Losses of soil carbon by converting tropical forest to plantations: erosion and decomposition estimated by δ13C. Glob Chang Biol 21:3548–3560

Harahap F, Silveira S, Khatiwada D (2017) Land allocation to meet sectoral goals in Indonesia: an analysis of policy coherence. Land Use Policy 61:451–465

Harsono SS, Grundmann P, Soebronto S (2014) Anaerobic treatment of palm oil mill effluents: potential contribution to net energy yield and reduction of greenhouse gas emissions from biodiesel production. J Clean Prod 64:619–627

Hess P et al (2009) Air quality issues associated with biofuel production and use. Biofuels: environmental consequences and interactions with changing land use. In: Howarth RW, Bringezu S (eds) Biofuels: environmental consequences and interactions with changing land use, proceedings of the scientific committee on problems of the environment (SCOPE) international biofuels project rapid assessment. Cornell University, Ithaca

Hewitt CN et al (2009) Nitrogen management is essential to prevent tropical oil palm plantations from causing ground-level ozone pollution. PNAS Proc Natl Acad Sci 106(44):18447–18451

IGES (2010) Is Indonesia in a good position to achieve sustainable low carbon development? Opportunities, potentials and limitations. IGES, Hayama

IPCC (2007) Climate change 2007: mitigation of climate change, Contribution of Working Group III to the fourth assessment report of the intergovernmental panel on climate change. Cambridge, Cambridge University Press

Jelsma I, Schoneveld GC, Zoomers A, van Westen ACM (2017) Unpacking Indonesia's independent oil palm smallholders: an actor-disaggregated approach to identifying environmental and social performance challenges. Land Use Policy 69:281–297

Jinn MC et al (2015) Empty fruit bunches in the race for energy, biochemical, and material industry. In: Hakeem K, Jawaid MY, Alothman O (eds) Agricultural biomass based potential materials. Springer, Berlin, pp 375–389

Koh LP, Ghazoul J (2010) Spatially explicit scenario analysis for reconciling agricultural expansion, forest protection, and carbon conservation in Indonesia. PNAS: Proc Natl Acad Sci 107(24):11140–11144

Koh LP, Wilcove DS (2008) Is oil palm agriculture really destroying tropical biodiversity? Conserv Lett 1(2):60–64

Koh LP, Miettinen J, Liew SC, Ghazoul J (2011) Remotely sensed evidence of tropical peat land conversion to oil palm. Proc Natl Acad Sci U S A 108:5127–5132

Lee JSH, Garcia-Ulloa J, Koh LP (2012) Biofuel expansion in Southeast Asia: biodiversity impacts and policy guidelines. In: Gasparatos A, Stromberg P (eds) Socioeconomic and environmental impacts of biofuels. Cambridge University Press, Cambridge, pp 231–254

Lee JSH, Ghazoul J, Obidzinski K, Koh LP (2014) Oil palm smallholder yields and incomes constrained by harvesting practices and type of smallholder management in Indonesia. Agron Sustain Dev 34:501–513

Li TM (2015) Social impacts of oil palm in Indonesia: a gendered perspective from West Kalimantan. Center for International Forestry Research (CIFOR), Bogor

MA (2005a) Millennium ecosystem assessment: current state and trends assessment. Island Press, Washington, DC

MA (2005b) Millennium ecosystem assessment: responses assessment. Island Press, Washington, DC

MA. (2005) Millennium ecosystem assessment: biodiversity synthesis. Island Press, Washington, DC

McCarthy JF (2010) Processes of inclusion and adverse incorporation: oil palm and agrarian change in Sumatra, Indonesia. J Peasant Stud 37:821–850

McCarthy JF, Gillespie P, Zen Z (2012a) Swimming upstream: local Indonesian production networks in globalized palm oil production. World Dev 40:555–569

McCarthy JF, Vel JAC, Afiff S (2012b) Trajectories of land acquisition and enclosure: development schemes, virtual land grabs, and green acquisitions in Indonesia's Outer Islands. J Peasant Stud 39:521–549

Mekhilef S, Siga S, Saidur R (2011) A review on palm oil biodiesel as a source of renewable fuel. Renew Sust Energ Rev 15:1937–1949

Mellko H (2008) Water footprint for biofuels for transport – Finland and the EU 2010. Master of science thesis Helsinki University of Technology, Helsinki

Menichetti E, Otto M (2009) Energy balance and greenhouse gas emissions of biofuels from a product life-cycle perspective. In: Howarth RW, Bringezu S (eds) Biofuels: environmental consequences and interactions with changing land use. Proceedings of the Scientific Committee on Problems of the Environment (SCOPE) International Biofuels Project Rapid Assessment, Cornell University, Ithaca

Meyer MA, Chand T, Priess JA (2015) Comparing bioenergy production sites in the Southeastern US regarding ecosystem service supply and demand. PLoS One 10(3):1–24. https://doi.org/10.1371/journal.pone.0116336

Moore S, Evans CD, Page SE, Garnett MH, Jones TG, Freeman C, Hooijer A, Wiltshire AJ, Limin SH, Gauci V (2013) Deep instability of deforested tropical peatlands revealed by fluvial organic carbon fluxes. Nature 493:660–663

Mukherjee I, Sovacool BK (2014) Palm oil-based biofuels and sustainability in southeast Asia: a review of Indonesia, Malaysia, and Thailand. In: Renewable and sustainable energy reviews, vol 37, pp 1–12

Mulyoutami E et al (2010) Chapter 2: component a: land use and human livelihoods. In: Tata HL, van Noordwijk M (eds) Human livelihoods, ecosystem services and the habitat of the Sumatran orangutan: rapid assessment in Batang Toru and Tripa. World Agroforestry Centre (ICRAF) Southeast Asia Regional Office, Bogor

Muyibi SA, Ambali AR, Eissa GS (2008) Development-induced water pollution in Malaysia: policy implications from an econometric analysis. Water Policy 10(2):193–206

NBPOL (2009) Sustainability report 2009. New Britain Palm Oil Limited

Nesadurai HES (2013) Food security, the palm oil-land conflict nexus, and sustainability: a governance role for a private multi-stakeholder regime like the RSPO? Pac Rev 26:505–529

Ng YG, Shamsul Bahri MT, Irwan Syah MY, Mori I, Hashim Z (2014) Ergonomics observation: harvesting tasks at oil palm plantation. J Occup Health 55:405–414

Nikoloyuk J, Burns TR, de Man R (2010) The promise and limitations of partnered governance: the case of sustainable palm oil. Corp Gov Int J Bus Soc 10:59–72

Obidzinski K, Andriani R, Komarudin H, Andrianto A (2012) Environmental and social impacts of oil palm plantations and their implications for biofuel production in Indonesia. Ecol Soc 17(25). https://doi.org/10.5751/ES-04775-170125

Obidzinski K, Dermawan A, Hadianto A (2014) Oil palm plantation investments in Indonesia's forest frontiers: limited economic multipliers and uncertain benefits for local communities. Environ Dev Sustain 16:1177–1196

OECD (2008) Biofuel support policies: an economic assessment. Organization for Economic Co-Operation and Development, Paris

Oosterveer P, Adjei BE, Vellema S, Slingerland M (2014) Global sustainability standards and food security: exploring unintended effects of voluntary certification in palm oil. Glob Food Secur 3:220–226

Panichelli L, Dauriat A, Gnansounou E (2009) Life cycle assessment of soybean-based biodiesel in Argentina for export. Int J Life Cycle Assess 14(2):144–159

Partzsch L (2009) The legitimacy of biofuel certification. Agric Hum Values 28:413–425

Pricewaterhouse Coopers (2010) Overview of palm oil industry landscape in Indonesia. In Palm oil plantation: industry landscape, regulatory and financial overview. Pricewaterhouse Coopers

Putrasari Y, Praptijanto A, Santoso WB, Lim O (2016) Resources, policy, and research activities of biofuel in Indonesia: a review. Energy Rep 2:237–245

Rahyla RS, Radin Firdaus RB, Purwaningrum F, Yildiz F (2017) Upgrading of Malaysian palm oil biofuel industry: lessons learned from the USA and Germany's policies. Cogent Food Agric 3:1279760

Ramdani F, Hino M (2013) Land use changes and GHG emissions from tropical forest conversion by oil palm plantations in Riau Province, Indonesia. PLoS One 8:e70323

Raymond CM et al (2009) Mapping community values for natural capital and ecosystem services. Ecol Econ 68(5):1301–1315

Reed MS (2008) Stakeholder participation for environmental management: a literature review. Biol Conserv 141(10):2417–2431

REN21 (2010) Renewables 2010 global status report. REN21, Paris

REN21 (2016) Renewables 2016 global status report. REN21, Paris

RFA (2008) The Gallagher review of the indirect effects of biofuels production. East Sussex, Renewable Fuels Agency

Rist L, Feintrenie L, Levang P (2010) The livelihood impacts of oil palm: smallholders in Indonesia. Biodivers Conserv 19:1009–1024

Robles MT, von Braun M, J. (2009) When speculation matters. International Food Policy Research Institute, Washington. DC

Romeu-Dalmau, C. Gasparatos, A., von Maltitz, G., Graham, A., Almagro-Garcia, J., Wilebore, B., Willis, K.J. (2017) Impacts of land use change due to biofuel crops on climate regulation services: five case studies in Malawi, Mozambique and Swaziland. Biomass Bioenergy (in press) https://doi.org/10.1016/j.biombioe.2016.05.011

RSPO (2007) Principles and criteria for the production of sustainable palm oil. Roundtable on Sustainable Palm Oil (RSPO), Kuala Lumpur

RSPO (2012) ACOP digest: a snapshot of RSPO members' annual communications of progress. Roundtable on Sustainable Palm Oil (RSPO), Kuala Lumpur

Runge CF, Senauer B (2008) How ethanol fuels the food crisis. Foreign Affairs

Savilaakso S, Garcia C, Garcia-Ulloa J, Ghazoul J, Groom M, Guariguata MR, Laumonier Y, Nasi R, Petrokofsky G, Snaddon J, Zrust M (2014) Systematic review of effects on biodiversity from oil palm production. Environ Evid 3:1–21

Sayer J, Ghazoul J, Nelson P, Klintuni Boedhihartono A (2012) Oil palm expansion transforms tropical landscapes and livelihoods. Glob Food Sec 1:114–119

Schoneveld G (2010) Potential land use competition from first-generation biofuel expansion in developing countries. Center for International Forestry Research (CIFOR), Bogor

Schouten G, Glasbergen P (2011) Creating legitimacy in global private governance: the case of the roundtable on sustainable palm oil. Ecol Econ 70:1891–1899

Sheil D, Casson A, Meijaard E, Noordwijk M, Gaskell J, Sunderland-Groves J et al (2009) The impacts and opportunities of oil palm in Southeast Asia: what do we know and what do we need to know? Center for International Forestry Research (CIFOR), Bogor

Silva-Castaneda L (2012) A forest of evidence: third-party certification and multiple forms of proof—a case study of oil palm plantations in Indonesia. Agric Hum Values 29:361–370

Stringer LC et al (2006) Unpacking "participation" in the adaptive management of social–ecological systems: a critical review. Ecol Soc 11(2):39

Stromberg P, Gasparatos A, Lee JSH, Garcia-Ulloa J, Koh LP, Takeuchi K (2010) Impact of biofuels on ecosystem services and biodiversity. (UNU-IAS Policy Report), UNU-IAS, Yokohama

Tata HL, van Noordwijck M, Muloyutami E, Rahayu S, Widayati A, Mulia R (2010) Human livelihoods, ecosystem services and the habitat of the Sumatran orangutan: rapid assessment in Batang Toru and Tripa. World Agroforestry Centre (ICRAF), Bogor

Tauli-Corpuz V, Tamang P (2007) Oil palm and other commercial tree plantations monocropping: impacts on indigenous peoples' land tenure and resource management systems and livelihoods. Forum on Indigenous Issues, Sixth session, New York, 14–25 May

TE (2016) Cars and trucks burn almost half of palm oil used in Europe. Transport and Environment, Brussels

TEEB (2010) The economics of ecosystems and biodiversity: ecological and economic foundations. Earthscan, London

UNICEF (2016) Palm oil and children in Indonesia: exploring the sector's impact on children's rights. UNICEF, Jakarta

USAID (2009) November 2008 – January 2009 Papua assessment USAID/Indonesia Final Report. USAID

Van der Werf GR et al (2008) Climate regulation of fire emissions and deforestation in equatorial Asia. PNAS: Proc Natl Acad Sci 105(51):20350–20355

Van Lienden AR et al (2010) Biofuel scenarios in a water perspective, changes in water footprints due to changes in the transport sector, Value of water research report series no.43. UNESCO-IHE, Delft

van Straaten O, Corre MD, Wolf K, Tchienkoua M, Cuellar E, Matthews RB, Veldkamp E, Vitousek PM (2015) Conversion of lowland tropical forests to tree cash crop plantations loses up to one-half of stored soil organic carbon. Proc Natl Acad Sci U S A 112:9956–9960

von Geibler J (2013) Market-based governance for sustainability in value chains: conditions for successful standard setting in the palm oil sector. J Clean Prod 56:39–43

White J, White B (2012) Gendered experiences of dispossession: oil palm expansion in a Dayak Hibun community in West Kalimantan. J Peasant Stud 39:995–1016

Wiggins S, Henley G, Keats S (2015) Competitive or complementary? Industrial crops and food security in Sub-Saharan Africa. ODI, London

Winrock (2009) Implications of biofuel sustainability standards for Indonesia. White paper #4, Winrock International

Woittiez LS, van Wijk MT, Slingerland M, van Noordwijk M, Giller KE (2017) Yield gaps in oil palm: a quantitative review of contributing factors. Eur J Agron 83:57–77

Woods J (2006) Science and technology options for harnessing bioenergy's potential. In: Hazell P, Pachauri RK (eds) Bioenergy and agriculture, promises and challenges. International Food Policy Research Institute (IFPRI), Washington, DC

Wu TY, Mohammad AW, Jahim JM, Anuar N (2010) Pollution control technologies for the treatment of palm oil mill effluent (POME) through end-of-pipe processes. J Environ Manag 91(7):1467–1490. https://doi.org/10.1016/j.jenvman.2010.02.008

WWF (2006) WWF Indonesia. Accessed at http://www.wwf.or.id/index.php?fuseaction=news.detail&language=e&id=NWS1158052088

Zah R et al (2008) Life cycle assessment of energy products: environmental impact assessment of biofuels. ETH – EMPA, St. Gallen

Zhou A, Thomson E (2009) The development of biofuels in Asia. Appl Energy 86(1):11–20

Part V
Sustainable Biofuels Strategy Options: Roadmap for Building Sustainable Strategy Options

Chapter 11
Roadmap for Building Sustainable Strategy Options

Masahiro Matsuura and Hideaki Shiroyama

11.1 Mutual-Gains Approach to Sustainable Policy

11.1.1 Failure of Command and Control Approach

Traditionally, the command and control approach has dominated the realm of environmental regulations. In a nutshell, the government is supposed to set a standard by obtaining objective scientific information and conducting a rational assessment of risks and benefits, and to enforce it by conventional stick and carrot mechanisms such as monitoring and penalty. In reality, however, this model has turned out to be not as effective as it was supposed to be. First of all, the cost of monitoring all regulated activities turns out to be too large for the public to pay for. While there have been efforts to improve monitoring devices, only a few who tries to make a large sum of short-term profit by evading regulations can do an enormous harm to the environment. Secondly, the command and control approach encouraged distrust among stakeholders. Supervising agencies and environmental groups are always being skeptical about what the industry does. Meanwhile, the industry becomes frustrated with the regulatory pressures and tried to manipulate through lobbing. In the end, rule-making processes become acrimonious, and the main goal of protecting the public through regulation is forgotten in the battle.

Thus, a new approach to rule-making is needed. In preparing the sustainable strategy options for biofuel utilization, an alternative to command and control is needed as well. It is simply impossible to set an ideal strategy for sustainable biofu-

M. Matsuura (✉)
Graduate School of Governance Studies, Meiji University,
1-1 Kanda Surugadai, Chiyoda-ku, Tokyo 101-8301, Japan
e-mail: mmatsuura@meiji.ac.jp

H. Shiroyama
Graduate School of Law and Politics, The University of Tokyo,
Tokyo, Japan

K. Takeuchi et al. (eds.), *Biofuels and Sustainability*, Science for Sustainable Societies, https://doi.org/10.1007/978-4-431-54895-9_11

els and hope to have it implemented by powerful leaders and government agencies in the current environment of politics.

11.1.2 An Alternative: Mutual-Gains Approach

While different alternatives to command and control approach have been discussed in the field of environmental policy, authors here focus on the mutual-gains approach. The basic premise of the approach is to foster voluntary agreement among stakeholders. As we see in the command and control approach, if each stakeholder tries to "win" an argument over its adversaries, then there won't be any cooperation among them. When everyone tries to "win" and expects others to "lose" (so-called "win-lose" situations), the result is often a "lose-lose" situation in which all stakeholders fare less than they could have achieved because they cannot create values through cooperation.

In order to achieve so-called "win-win" outcomes, each party must be willing to cooperate with other parties to create values by exploring the areas that it can contribute to the other side. Toward this goal, different stakeholders, including the government, industry, and NGOs, have to negotiate on the equal footings. Government agencies are not endowed with the lightening rod any more.

In the context of biofuel utilization strategy, contributors to this volume have identified a wide variety of stakeholders. While government agencies are one of the key actors, there are many others who have the significant power in the course of the implementation of biofuel utilization even in the developing nations. Therefore, an alternative approach to strategy building that seeks voluntary agreements among the stakeholders is likely to produce more effective strategies than other approaches that seek a realization of an ideal world through command and control and political struggles.

11.1.3 Practice of Mutual-Gains Policy Formulation

There have been a few practical advices regarding how stakeholders can find mutually beneficial agreements that they can live with. The first principle is to focus on interests, not positions (Fisher and Ury 1987). Stakeholders in biofuel utilization will unavoidably make positional statements in the course of strategy building, particularly if they are in a bad relationship in which each of them seeks a "win-lose" outcome. In many cases, however, such positional statements are exaggerated and a manifestation of their ideals that they hope to achieve. On the other hand, they have specific interests in the issue and hope to improve the situation. Interests are possible answers to "Why do you want the conditions that you made in the statement?" For instance, an environmental advocate might say, "No tree shall be cut!" If s/he is asked for why, s/he might say, "I'm concerned about the life of pristine orangutans!" It might be difficult for other stakeholders to accommodate the first claim, but maybe able to deal with the second one. Thus, understanding the interests behind positional statements can lead stakeholders to a productive negotiation that could lead to a mutually agreeable solution. This is the first principle of the mutual-gains policy formulation.

The second principle is a step ahead from the first principle. Once varying interests of stakeholders are identified, there might be possible trades between these multiple interests that can bring about benefits to both parties. In the case of orangutans, the developer might be able to try all possible measures to protect the forest where orangutans live, while the environmental advocate might be willing to concede in the development in the area where no orangutans live. Such trade is impossible if both parties insist on their rights and positional statements. Mutual-gains approach to policy formulation seeks such trade that brings about benefits to all stakeholders.

Lastly, any strategy building effort should recognize the bottom line of each stakeholder. The best alternative to a negotiated agreement, abbreviated as BATNA, is a condition that each stakeholding party decides to leave a voluntary agreement and take a unilateral action. Any collaborative strategy for policy formulation should provide each stakeholder a benefit whose size excels their BATNA. If the strategy is based on a wide variety of stakeholder interests, it would be able to provide sufficient benefits to each party because cooperative mechanisms embedded in the strategy can produce enough benefit for the stakeholders to share.

When some stakeholders have extreme expectations regarding their BATNA (e.g., they believe that any form of collaboration with other parties would harm their political interests), then mutual-gains approach to strategy building might not be possible to involve such stakeholders. In such cases, other stakeholders should probably give up such fundamentalists after trying to persuade the possible benefit of collaboration.

11.1.4 Challenges in Mutual-Gains Policy Formulation

While mutual-gains approach is likely to be more stable and efficient in the long run, compared to the traditional command and control approach, there are many challenges that the strategists have to be prepared to deal with. The first challenge, especially in the context of biofuel utilization, is the range of stakeholders that a strategy has to care about. As mentioned in the previous section, biofuel utilization strategy at the global, regional/national, and local level has to consider interaction with other levels of governance. For instance, local deployment strategy still has to consider the implication of sustainability standards, which is discussed at the global level, to the strategy. It also has to consider national policy and regulation as well.

Therefore, stakeholder-based approach entails difficulties with defining boundaries around the analysis. In practice, it is likely that there are a practical number of stakeholder representatives for each project and strategy building effort. The number of stakeholders involved in a project has to be in a manageable order. One pragmatic solution is to limit the number of stakeholders, while allowing other stakeholders to observe the progress of strategy building and provide meaningful inputs to the process. There is no theoretically defendable answer, however, as to the boundary of stakeholders to be considered in the strategy.

The second challenge is uncertainty regarding the implementation of the strategy. While mutual-gains approach is more resilient in this aspect than the inflexible command and control approach, shifting political environment might block the implementation of formerly agreed arrangement. For instance, a new president might be elected on a totally different platform on biofuels. In such instance, the strategy has to be revisited, and a new round of negotiation between stakeholders has to be organized. In addition to political uncertainties, there are also scientific uncertainties. We still do not know exactly what would happen if the concentration of greenhouse gases continues to increase, for example. We do not know what kind of innovations might occur in the future in the field of biofuels. In order to deal with such scientific uncertainties, the strategy has to have an embedded system that would allow periodical redesign of the strategy for incorporating the up-to-date scientific knowledge and innovations.

The last challenge is the tension between creating and claiming value. Theories of negotiation found the innate difficulties of bringing parties together for collaboration because the collaboration always has an aspect of competition (Lax and Sebenius 1986). In the context of biofuel strategy building, some stakeholders might hold on to their positional statements in the hope of obtaining more concessions from other parties. This is an inevitable challenge in implementing the mutual-gains approach to strategy building and has to be dealt with by professional process managers who have expertise in managing the tension among stakeholders in similar situations.

11.1.5 Mutual-Gains Biofuel Policy-Making in Action

There are a few examples of such mutual-gains approach identified in this volume.

The utilization of bagasse for the electricity generation, described in Sect. 2.1.2, is an interesting example of mutual-gains approach by involving different stakeholder groups in the picture of plantation development. While it would be difficult to justify the environmental impact of sugarcane-based ethanol production from the viewpoint of life cycle assessment, the same project can be justified by involving the interest of utility companies and electricity users who have concerns about the CO2 emissions from additional coal fire power plants. Electricity generation from bagasse is a typical but an ideal "win-win" solution that brings about benefits to all involved parties. The most interesting aspect of this case study is that, however, this predictable collaboration is now supported by a detailed study of life cycle assessment. This seems to provide evidence that the mutual-gains approach is in fact economically more efficient than conventional approaches encourage each stakeholder to focus only on their preconceived interests.

Community-based utilization of *Jatropha* in Indonesia, described in Sect. 2.2.2., is another interesting case of collaborative strategy building. The traditional top-down approach by the central Indonesian government to propagate the *Jatropha* production across the country has obviously failed few years after its inception in the early 2000s. The authors of the case study suggest an alternative that looks at the common interests of the local stakeholders who need basic fuel sources for cooking and other household jobs. Thus they suggest the use of *Jatropha* at the community scale. This community-based strategy might seem to have miniscule impacts compared to the national strategy. But the readers should be aware that the national strategy simply failed because it didn't address the interests of the stakeholders at the local level. The size of resources wasted in the national effort should not be left unnoticed. While the proposal for community-based *Jatropha* utilization might be a small contribution to the biofuel strategy in terms of the size, it is much less likely to produce the negative effect that the national program had in the past.

At the international level, RSPO, RSB, GBEP, and other organizations' effort for sustainability standard setting and other kinds of activities for improving the sustainability of biofuels are typical examples of mutual-gains approach. As mentioned in Sect. 1.2., these organizations explicitly cares about the attention to the full range of stakeholders related to the sustainability of biofuels. While they vary in terms of the scope of the issues and the approach to sustainability, their strategy seem robust in principle from the standpoint of mutual-gains approach. One concern would be, however, the involvement of full range of stakeholders and political processes within each organization. Operation of these organizations should be studied in details from the stakeholder perspectives further in the future.

11.2 Deliberative Policy Formulation for an Improved Sustainability

11.2.1 Concerns About Conventional Neoliberal Approaches

While mutual-gains approach to policy formulation seems to have an advantage over the traditional command and control approaches regarding the stability and predictability of implementation because of stakeholder supports, there have been a few sharp critiques on the way it has been manipulated by certain categories of stakeholders who has the power. In particular, mutual-gains approach assumes that a theme of the policy discussion is given by the stakeholder, or convenor, who initiates the policy-making effort. Those who have the power and resources to design the policy formulation process can manipulate the process quite easily in the name of public participation. For those who are concerned about the democratic nature of public policy processes, mutual-gains approach might not be paying enough attention to the power imbalance in the phase of agenda setting (Kingdon 1998).

The most common critique would be about the validity of stakeholder representatives in the forum of discussion. For instance, can we discuss sustainable biofuel utilization without involving the representatives of indigenous people who lives in the tropic forests of Indonesia? Some might argue that such stakeholders are represented by certain civil society organizations. Others might still criticize the representatives are "brived" by the organizer and thus their participation is not considered as a valid form of stakeholder representation.

In this context, the mutual-gains process could be manipulated in a way that reinforces power imbalance in policy-making processes between the rich and the poor. This kind of discussion can easily lead to the debate over the "fair" and "equitable" division of wealth created through collaborative efforts by stakeholders who have different power in the conventional processes.

These critiques do not completely dismiss the value of mutual-gains policy formulation processes per se. Rather, these are a kind of mild warming for us about the possible manipulations of processes by a limited number of powerful stakeholders. Anyone who organizes the policy formulation process is morally obliged to consider the "fair" processes regarding the choice of stakeholders and agenda. If the subject matter is related to the rights and value questions that cannot be resolved by focusing on the interests, one may consider taking a different path that primarily focuses on the deliberative aspects of policy discussions, as discussed below.

11.2.2 New Forms of Governance

Responding to such critiques, a new school of political scientists since the beginning of this century has started to explore a concept called deliberative democracy. In this framework of policy formulation, participants are asked to engage in a discussion as free citizens without worries about the value creation and other self-interests. Instead, they engage in discussions based on "reasons" and try to identify a common set of ideas they can agree with irrespective of their own interests.

The idea for deliberation, drawing on Greek tradition of political debate but recently revitalized by Gutmann and Thompson (1998), tries to address moral questions that cannot be solved though bargaining over individual interests that is presupposed by the mutual-gains thinking. For instance, one may question what the "sustainability" means. This is not a matter of discussion of bargaining. It is more about the public perception and theoretical discussion about what the public accepts as a norm and common language.

Practitioners, particular in the field of science policy, have explored the application of deliberative discussions. For instance, Danish office of technology assessment has been gathering members of the public by random sampling and asking them for a deliberation over important scientific issues. Other kinds of deliberative democracy projects have been experimented in northern European countries, as well as in some parts of the United States.

11.3 Resilient Governance

11.3.1 Uncertainty and Governance

High levels of uncertainty require a different strategy formulation process that pays particular attention to its risk and benefit. The command conventional and control approach fares the worst in such environments. It assumes government agency's unilateral imposition of previously determined regulations, which has undergone rigorous examination of the public decision-making processes. Whenever the environment surrounding the regulation changes, the government agency has to revisit the configuration of regulations by conducting a "rational" analysis, propose an alternative set of regulations, and go through the rigorous (and often time consuming) public decision-making processes. Such closed and stringent systems cannot fit with the rapidly changing environment, leading everyone into a terrible situation.

Mutual-gains policy formulation and other kinds of deliberative processes, however, can also be even more time consuming especially if they have to do the discussion from the scratch every time the situation changes.

Under the high level of uncertainties, it would be quite difficult for the stakeholders to come up with a comprehensive agreement because there are so many question marks regarding what might happen in the near future. For instance, how far can we be confident that there will be no severe weather conditions that can harm the production of feedstock? We might know how likely it is, but we can't no definitely whether it will happen or not in a foreseeable future. Do we know exactly when new robust innovations for biofuels production will be available? It is advisable for the stakeholder group to stay away from debating over these questions because we simply don't know when it really happens.

An alternative is to shift the focus from decision-making processes to institutional developments while maintaining the principles of mutual-gains and/or deliberative discussions. Under the high level of uncertainty, strategy does not have to be finalized, but the working group of stakeholders and/or selected members of the public needs to be set up so that they can reconvene quickly and periodically after new information or situation comes up.

This means a creation of institutional mechanism for dealing with the ever-changing situations. The mechanism must be structured as an open system that allows flexible reconfigurations of participants and agenda in order to avoid the capture of the process by a few powerful interest groups.

11.3.2 *Creating Resilient Institutional Mechanisms for Biofuel Utilization*

How could we incorporate the argument for resilient governance into the discussion of biofuel utilization strategy? The question has to be answered for different levels of governance.

At the local level, biofuel project might be better conceived as an institutional building rather than as a project that completes within a specific time frame. Through the mutual-gains model, they might be able to reach a mutually satisfactory utilization strategy. They might be able to deploy a conventional technology in a short run with satisfaction to every stakeholder. In addition to that once-through process, they are encouraged to form an organization and institutionalize rules regarding how they maintain and reconfigure the project outcomes. A new technology might be available only 1 year after the completion of the project. Local weather condition might change due to climate change, and the necessary feedstock might become unavailable in the field. With such institutionalized mechanisms, local stakeholder can easily adapt its biofuel utilization strategy to the changing environment.

At the regional/national level, the same kind of organization is necessary to deliberate and negotiate on the biofuel policy. Such a body has to set forth biofuel policy and regulations in a timely manner. This regional/national arrangement has to be in accordance with the global and the local level.

Lastly, at the global level, institutional development has already begun by a few stakeholder-focused initiatives, such as RSPO and RSB. While they produce sustainability standards and other kinds of guidelines as a product of their mutual-gains policy formulations, the organization itself is a manifestation of institutional development (i.e., these organizations are not disbanded after the preparation of certain documents). A wide variety of stakeholders continuously collaborate under these institutions. One possible concern about these institutions hinges on their openness. Are they willing to change its membership according to the changing situations in the field? Do they engage in the reflective practice that periodically questions the effectiveness of institutional arrangement? Detailed studies on the actual management of such international organizations are much needed.

Chapter 12
Application of Ontology for Developing Strategy Options

Kouji Kozaki, Osamu Saito, Masahiro Matsuura, and Riichiro Mizoguchi

12.1 Introduction

One of the core questions for sustainability science is investigating how the dynamic interactions between nature and society can be better incorporated into emerging models and conceptualizations that integrate the Earth system, social system, and human system (Kates et al. 2001; Komiyama and Takeuchi 2006). Since these interactions, by their nature, relate to various stakeholders and players from many different fields, the problem-solving process requires the collaboration and partnership of these players. Many efforts have been made to structure diverse and fragmented knowledge for facilitating their collaboration (Choucri et al. 2007; Kumazawa et al. 2009).

K. Kozaki (✉)
Department of Knowledge Science, The Institute of Scientific and Industrial Research (ISIR), Osaka University, 8-1 Mihogaoka, Ibaraki, Osaka 567-0047, Japan
e-mail: kozaki@ei.sanken.osaka-u.ac.jp

O. Saito
United Nations University Institute for the Advanced Study of Sustainability (UNU-IAS), 5-53-70 Jingumae, Shibuya 150-8925, Tokyo, Japan

M. Matsuura
Graduate School of Governance Studies, Meiji University, 1-1 Kanda Surugadai, Chiyoda-ku, Tokyo 101-8301, Japan

R. Mizoguchi
Research Center for Service Science, Japan Advanced Institute of Science and Technology, 1-1 Asahidai, Nomi 923-1292, Ishikawa, Japan

K. Takeuchi et al. (eds.), *Biofuels and Sustainability*, Science for Sustainable Societies, https://doi.org/10.1007/978-4-431-54895-9_12

Consensus-building among various stakeholders from different fields is one of key issues to solve for facilitating their collaboration. In order to build consensus, it is important to know what others are thinking about each other because differences of their viewpoints cause some conflicts. However, it is difficult to understand different views in particular when they come from different fields. To overcome this problem, we took an ontology-based approach.

Gruber (1993) defined ontology as an "explicit specification of conceptualization." A well-constructed ontology can present an explicit essential understanding of the target world. Based on ontology engineering, a wide range of knowledge can be organized in terms of general, highly versatile concepts and relationships. In order to provide a base knowledge for consensus building across various domains, the authors have developed a biofuel ontology on the basis of the sustainability science ontology (Kumazawa et al. 2009), literature surveys, and stakeholder analysis. And the authors have developed a divergent ontology exploration tool that can generate comprehensive conceptual maps from user's multiple arbitrary perspectives (Kozaki et al. 2011). The exploration tool allows the user to explore ontologies interactively according to their interests. The results of their explorations are visualized as conceptual maps. That is, the conceptual maps represent viewpoints of the users.

This section describes detail design and functions of ontology-based application system which supports consensus-building system based on the ontology exploration and effectiveness of ontology system for developing for biofuel strategy options.

12.2 System Architecture and Process

Chapter 3 introduces stakeholder perspectives and emphasizes the importance of multilevel governance. The purpose of stakeholder analysis is to indicate whose interests should be taken into account and why they should be taken into account during decision-making process on a particular issue (Crosby 1991). This analysis also focuses on the quantity and types of resources those groups or actors can mobilize to affect outcomes regarding that issue. Stakeholder analysis encompasses a range of different methodologies and tools for analyzing stakeholder interests. This analysis should be generally conducted by an independent researcher/organization viewed as neutral to the issue in focus (Fig. 12.1).

On the other hand, this chapter explains the ontology-based knowledge structuring and visualizing (mapping) system that can facilitate holistic framing and collaboration among various stakeholders in a particular issue. By using this system, users (stakeholders) can explore various conceptual linkages regarding their specific interests and create conceptual maps which visualize relevant concepts with semantic links (nodes) around the focal concept (Fig.12.1).

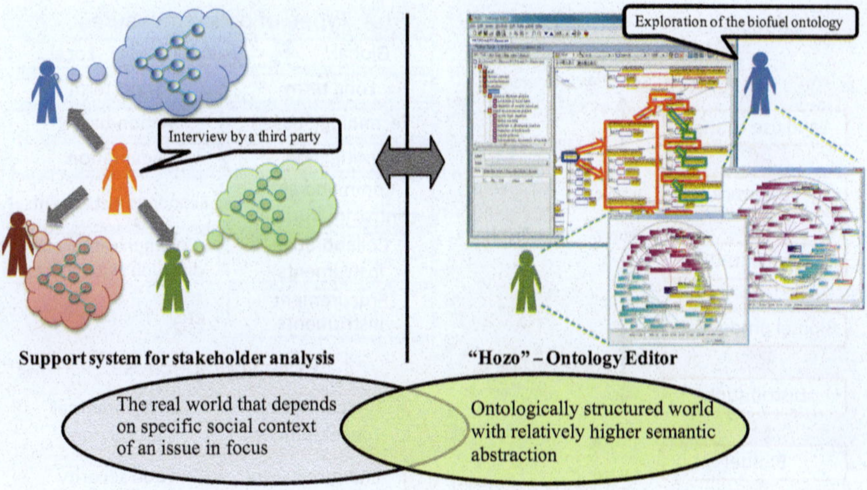

Fig. 12.1 Collaboration between stakeholder analysis and ontology engineering

Through our research project on sustainable biofuel, we argued how to apply ontology engineering to stakeholder analysis and enhance function of the existing ontology-based system to support stakeholder analysis. For this purpose, the gap between the two approaches was identified: stakeholder analysis treats concrete world that depends on specific social context of an issue in focus, while ontology engineering emphasizes structured world with relatively higher semantic abstraction. Then, modification and function enhancement were made to bridge the gap. For example, the existing biofuel ontology was extensively upgraded on the basis of research outcomes by stakeholder analysis. The system interface and functions were also improved to enable multiple users (stakeholders) to use the system at the same time during the decision-making process.

Based on the stakeholder analysis in Chaps. 3 and 9, we can identify four different dimensions for planning biofuel policy measures (Fig. 12.2). The first one is the life cycle of biofuel from land use change by energy crop cultivation, biofuel production, distribution, and endues of biofuel. Stakeholders are second dimension which often includes various players in both developed and developing countries. Types of policy measures as third dimension consider if a policy should or can be applied to global, regional, or local scale and if it is long term or short term, technology-based or action-based, and so forth. Fourth dimension asks from which perspective or objective a policy is designed. Economic development, energy security, food security, or water security, for example, would be one of those perspectives. Implemented and proposed policy measures were sorted out to meet these dimensions and integrated into the biofuel ontology.

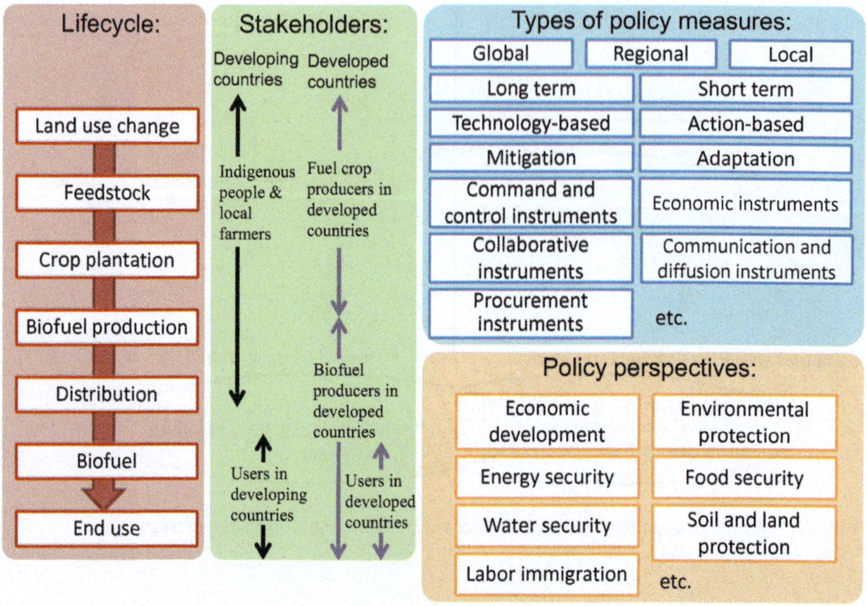

Fig. 12.2 Four dimensions for planning biofuel policy measures

12.3 Facilitation of Planning and Collaboration

Figure 12.3 shows the block diagram of the system for facilitating planning ad consensus building. The system behavior is composed of two steps. In the first step, each user (stakeholder) is asked to build a map based on his/her own interest. Collaborative work and/or discussion among them using the maps they generated is done in the second step. The interface of the system is designed to lighten the load of use of its functions to enable users to easily generate maps. The interaction with the system is interactive exploiting the current user-friendly technology such as tablet PCs and multitouch tables. Map visualization after exploring the ontology is easily done as well as post-editing of the map to make it compact and informative enough. Especially, easy interpretation of maps is essential for our research. To achieve this, a couple of useful functions for highlighting focused items in the map are prepared. For example, the target items include kinds of relations and concepts and perspectives such as global/local and long-term/short-term. "Change-view" function can redraw the map according to the specified item by the users to make the map more informative.

Figure 12.4 shows the map generated intended to extract the influence of the increase of biofuel production on the land use from the point of view of an environmental NGO. This map was generated by search path from "biofuel production" to "land use." Because the system takes account of all relationships related to not only the selected concepts but also subclasses of them, we can see many concepts related to them such as "forest area," "open burning," "area definition problem of farm land uti-

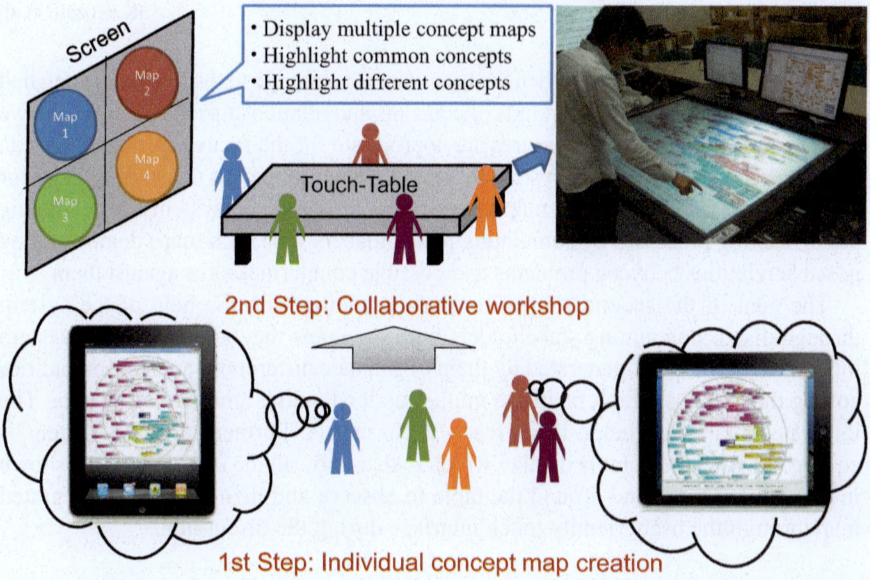

Fig. 12.3 System design of the planning and consensus building facilitation

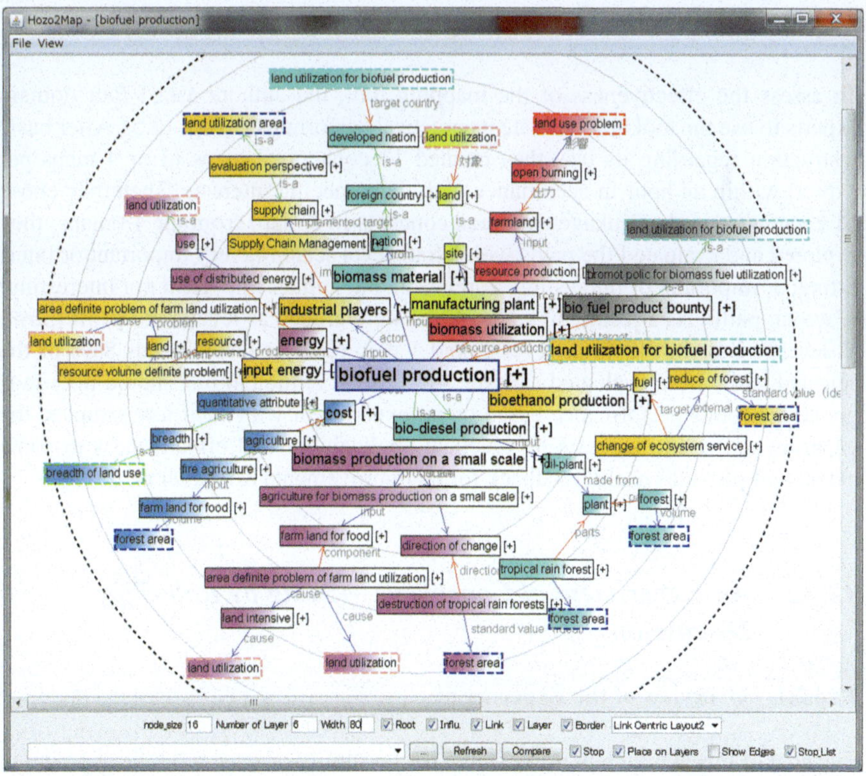

Fig. 12.4 An example of conceptual map generated from the point of view of an environmental NGO

lization," etc. from this map. When the user wants to generate maps from more detailed viewpoints, he/she can specify kinds of concepts and relationships to follow. When we want to know what countermeasures are appropriate for the focused problem, we can obtain another map using the system by selecting the problem as the starting point for an ontology exploration. This map suggests the utility of the system for facilitating policy making processes by stimulating policy makers with such maps demonstrating possible relations between problems and possible countermeasures against them.

The goal of the second step is consensus making with the help of the system through discussion among stakeholders with the maps they generated. The system integrates all the maps generated by them to enhance differences and commonalities among those maps which facilitate mutual understanding among participants. The integrated map thus helps them reach a consensus. Furthermore, the system is equipped with a touch table display which is shared by all the stakeholders as shown in Fig. 12.3. They stand around the table to observe and manipulate the integrated map through the user-friendly touch interface during the discussion.

12.4 Usability and Effectiveness of the System

12.4.1 Evaluation Experiment by Domain Experts

To assess the effectiveness of the mapping tool, the authors asked four domain experts to use the tool and evaluate its practical performance (Fig. 12.5). After basic instruction regarding its use, they created 13 conceptual maps (3 or 4 maps per expert) within an hour in accordance with their specific interests. Then they chose 61 conceptual paths (linkages between concepts in a map) from the 13 maps; they explored and evaluated the paths with a four-level scale (4, very important or interesting; 3, important or interesting; 2, relevant, but neither important nor interesting; 1, wrong path). As a result, 30 paths (49%) were graded as level 4, 22 paths (36%) as level 3, 8 paths (13%) as level 2, and 1 path (2%) as level 1; thus 85% of the selected paths were evaluated as level 3 or level 4. Although one should not exaggerate the tool's performance based on an experiment with such few samples, the experimental result suggests its practical applicability and effectiveness to some extent and provides useful feedback for its improvement (Kozaki et al. 2011).

12.4.2 An Experiment of Consensus Making by Role-Play Discussion

12.4.2.1 Overview of the Experiment

The goal of this experiment is to explore the feasibility of system. In the experiment we assigned a couple of subjects roles of stakeholders related to biofuel production and policy making for it and ask them to discuss the related topics by role-playing

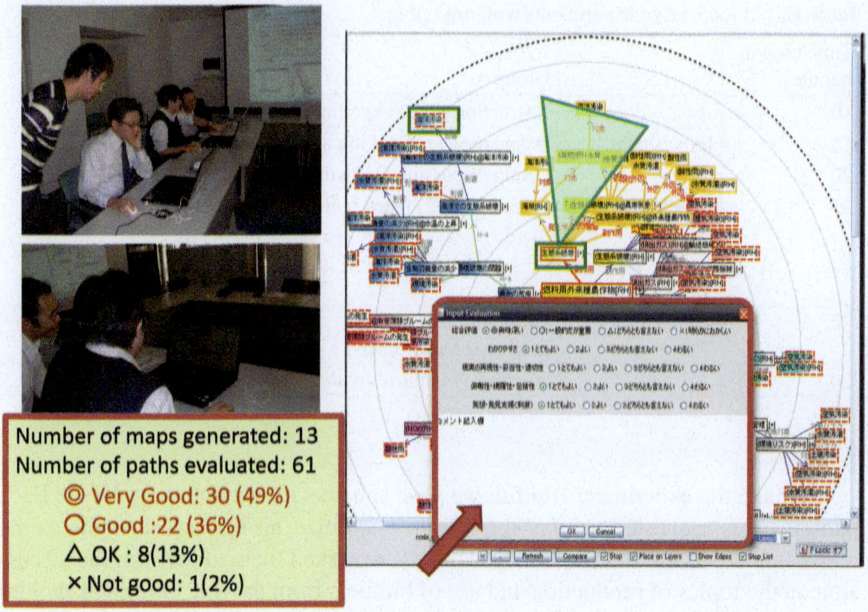

Fig. 12.5 Experimental expert workshop for application and evaluation of the tool

and to explore the possibility to come to a better mutual understanding which would help them reach a reasonable consensus.

The subjects are composed of two junior students and two master course students in the department of Sustainable Energy and Environmental Engineering of Faculty of Engineering (group A). In addition to them, we invited four researchers in the sustainability science domain (group B). Another researcher in the sustainability science domain joined in the discussion done among group A to coordinate the discussion.

12.4.2.2 Methods

Table 12.1 shows the detail of the experiment with time table. Group A conducted two discussions: one without the system (experiment 1) and the other with it (experiment 2). Group B also did two discussions but neither used the system. After the experiments, we also discussed the utility and usability of the system.

The roles of stake holders used in the experiment are as follows:

(a) Industry (sugarcane farmers, investors, sugar processing/brewery plants, etc.)
(b) Government (president, the relevant ministry, etc.)
(c) Employees (labor unions, etc.)
(d) Environmental NGO

Table 12.1 Processes of experiments with time table

<table>
<tr><th colspan="2">Time used in minute</th><th></th><th>Group A</th><th>Group B</th></tr>
<tr><td colspan="2">10</td><td></td><td colspan="2">Instruction of the experiment</td></tr>
<tr><td colspan="2">15</td><td rowspan="2">Experiment 1</td><td colspan="2">Preparation (1) [making a rough plan]</td></tr>
<tr><td colspan="2">20</td><td colspan="2">Group discussion (1) [without the system]</td></tr>
<tr><td rowspan="2">35</td><td>15</td><td rowspan="3">Experiment 2</td><td rowspan="2">Preparation (2) [each builds a map]</td><td>Preparation (2) [rough planning]</td></tr>
<tr><td>20</td><td>Group discussion (2) [without a map]</td></tr>
<tr><td colspan="2">20</td><td>Group discussion (2) [discussion with maps]</td><td>Participate in the discussion by group B</td></tr>
<tr><td colspan="2">20</td><td></td><td colspan="2">Answering inquiries with wrap-up discussion</td></tr>
</table>

To make the experiment fruitful, we gave subjects instructions as follows: Each participant is requested to play the role to maximize his/her own benefits as the representative of the stakeholder. Concretely, we asked them to perform the discussion on the topics of production and use of biofuels from the role of the stakeholder with the following items in their mind:

- Negative opinions: problems to be solved and anything needs improvement, etc.
- Positive opinions: what you expect, what you utilize, etc.

We also asked them to summarize the discussion on the following items in a summary sheet:

- In what respects your opinion conflicts with others'
- Other stakeholders with which you can collaborate on what respects

In the experiment 2 of group A with maps, each subject built a map after a brief instruction on how to use the system. The focal point from which exploration is done was set to "production of biofuels," and each subject built a map selecting a couple of keywords (3–5) from about 120 keywords prepared in advance. To minimize the deviation of the generated maps, we restrict the map generation command to "search path" which generates a map automatically according to the selected keywords. To make the maps compact and easy to interpret, we asked them to delete paths which they find not interesting and to extend such paths that they want to explore further. By doing this, they got maps including only interesting and meaningful paths from the perspective of the stakeholder role they play.

The subjects performed the discuss using the integrated map presented on the touch table with appropriate enhancement of interested items to contrast differences and commonalities among maps they made based on their own perspectives (Fig. 12.6). They thus exchange opinions with such a help provided by the system.

Fig. 12.6 A snapshot of the discussion around the touch table

Table 12.2 Number of nodes and overlapping nodes

	Number of nodes in the map	Number of overlapping nodes			
		(a) Industry	(b) Government	(c) Employees	(d) Environmental NGO
(a) Industry	110	–	16	21	10
(b) Government	88	16	–	12	5
(c) Employees	187	21	12	–	49
(d) Environmental NGO	115	10	5	49	–

12.4.2.3 Results and Discussion

Table 12.2 shows the number of nodes included in each map built by each subject in group A and those of the overlapping nodes between them. The numbers of overlapping nodes indicate the how much the stakeholders share common interests (Fig. 12.7). Comparison between these numbers reveals that employees and environmental NGO share a lot of common interests. This interpretation is supported by the fact that both employees and environmental NGO are classified into the same category citizen in the result of stakeholder analysis (Shiroyama et al. 2010). We believe such a function that derives quantitative information between stakeholders is one of the merits of the system. In addition to this, we found a couple of results which show particular relations between stakeholders which we did not expected before.

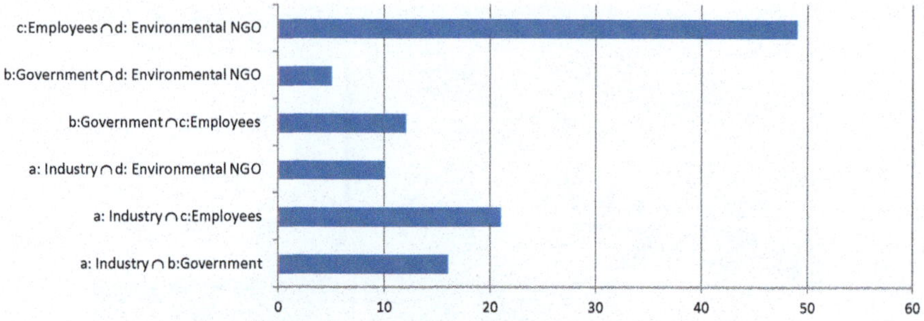

Fig. 12.7 Number of overlapping nodes between maps of stakeholders

The positive opinions we got from the subjects include:

- Visualization of conceptual maps is helpful to understand in what respects we are different by identifying what concepts we share and don't from the map.
- It sometimes helps us to realize the issues better by explicating unexpected relations or dependencies between concepts.
- It is useful for organizing my opinion to enable smooth discussion.
- It is useful to reveal overlap and distinction between us objectively.

These show the feasibility and utility of the system to some extent.

Comparison between the discussion done by groups A and B shows something interesting. While there is no significant differences of number of utterances between them, the number of utterances appearing the second discussion done by group A is significantly smaller than that of the second discussions done by group B. This was partly because the subjects in group A took much time to learn how to use the system so that they did not have enough time to perform discussion. In fact, we had quite a few requests on improvement of the mapping tool. Furthermore, we found the discussion done by group B which includes quite a few concepts that are not covered by the current ontology. These facts suggest the system needs further improvement on its usability and extension of the ontology to cover wider and deeper topics. We plan to implement these modifications of the system to realize a useful and usable system for facilitating consensus making for policy making of biofuel production and utilization.

12.5 Conclusion

In this section, we proposed a consensus-building supporting system based on ontology exploration. The system generates conceptual maps through ontology exploration by the users. Because the generated maps represent the users' viewpoints to understand the target domains of the ontology, it could show differences of viewpoints through comparisons of them. In order to evaluate the system, we made an experiment of consensus building by role-play discussion in biofuel domain. The result shows an integrated map could well represent different viewpoints of several stakeholders and could help their consensus building through discussions using the map. It would contribute to consensus building and policy making on interdisciplinary domains which consist various fields across multiple domains.

The client application version of ontology exploration tool is implemented as an extended function of Hozo which is published as free software at http://www.hozo.jp. The prototype of its web service version, which only supports search path function, is also available at http://env-ss.hozo.jp/.

References

Crosby BL (1991) Stakeholder analysis: a vital tool for strategic managers, technical notes: a publication of USAID's implementing policy change project, No.2, 1–6

Choucri N, Mistree D, Haghseta F, Mezher T, Baker WR, Ortiz CI (eds) (2007) Mapping sustainability: knowledge e-networking and the value chain. Springer, New York

Gruber TR (1993) A translation approach to portable ontology specifications. Knowl Acquis 5(2):199–220

Kates RW, Clark WC et al (2001) Environment and development: sustainability science. Science 292(5517):641–642

Komiyama H, Takeuchi K (2006) Sustainability science: building a new discipline. Sustain Sci 1:1–6

Kozaki K, Hirota T, Mizoguchi R (2011) Understanding an ontology through divergent exploration. Proc ESWC. 2011, Part I, LNCS 6643:305–320

Kumazawa T, Saito O, Kozaki K, Matsui T, Mizoguchi R (2009) Toward knowledge structuring of sustainability science based on ontology engineering. Sustain Sci 4:99–116

Shiroyama H et al (2010) FY2009 Final Report for Hc-082 Biofuel Use Strategies for Sustainable Development (2) Formation of biofuel utilization strategies for sustainable development

Part VI
Sustainable Biofuels Strategy Options: Key Strategies for Policy Makers

Chapter 13
Key Strategies for Policymakers

Shinichi Arai and Hirotaka Matsuda

13.1 Strategy Options at the Global Level

13.1.1 Background for Sustainable Deployment Strategies and Response Measures

Biofuels[1] have been identified as having diverse environmental, social, and economic impacts, as discussed in Chap. 2. For this reason, the use of biofuels to realize a sustainable society requires study that takes into account the respective characteristics of biofuel deployment on a global, regional, national, and local scale. In this chapter, we examine deployment strategies for sustainable biofuels on a global scale by surveying the current and future issues that need to be considered. These issues include environmental impacts starting with the reduction of greenhouse gas (GHG) emissions from biofuels and including other issues such as energy security, food security, rural development, agriculture and industrial policy, trade, and north-south issues. We then examine ways that biofuel deployment strategies can address these issues to realize a sustainable society.

[1] In this chapter, "biofuels" refer to liquid fuels such as bioethanol and biodiesel, including second-generation biofuels such as those derived by decomposing cellulose. Fuelwood and other biomass and gas recovery through conventional means such as methane fermentation of waste are excluded. However, "bioenergy" includes conventional biomass energy such as from fuelwood.

S. Arai
Graduate School of Environmental Science, Hokkaido University,
N10W5, Sapporo 060-0810, Hokkaido, Japan

H. Matsuda (✉)
Graduate Program in Sustainability Science – Global Leadership Initiative (GPSS-GLI),
Graduate School of Frontier Sciences/Integrated Research System for Sustainability Science (IR3S), Institutes for Advanced Study (UTIAS), The University of Tokyo,
7-3-1, Hongo, Bunkyo-ku, Tokyo 113-8654, Japan
e-mail: matsuda@k.u-tokyo.ac.jp

K. Takeuchi et al. (eds.), *Biofuels and Sustainability*, Science for Sustainable Societies, https://doi.org/10.1007/978-4-431-54895-9_13

13.1.1.1 Biofuels and Their Environmental Impact

Biofuels are carbon neutral at the usage stage and, because they are renewable, can be studied as a means of addressing global warming. However, there are many items that need to be considered including changes in land use, encompassing forest conservation, as well as impacts on air, water, and soil quality and impacts on water resources, ecosystems, and biodiversity. It is also important to assess the indirect impact on land use. For these reasons, biofuels have been studied by the Scientific Committee on Problems of the Environment (SCOPE) of International Council for Science (ICS) as a worldwide body and by the USA and other countries (SCOPE 2008; UDAID 2009; USEPA 2011), in addition to existing study by the United Nations organizations cited later in this chapter. In order to assess these items appropriately, it is necessary to study deployment strategies from a global perspective.

The United Nations Environmental Programme (UNEP 2009) reports that the GHG balance in LCA for biofuels varies widely depending on the raw materials, biofuel generation technology, and methodological assumptions. For example, bioethanol from sugarcane can reduce GHG emissions by between 70% and 140% compared with gasoline, whereas corn can reduce GHG emissions as much as 60% but may also increase them by as much as 5%. Biodiesel from palm oil can reduce GHG emissions by as much as 80%, but when palm oil is harvested by converting natural forests into plantations, it can increase GHG emissions by 870–2000% when taking into account the impact of the land use change (UNEP 2009). In other words, biofuels do not always have the effect of reducing GHG emissions when compared with fossil fuels such as gasoline and diesel oil, if we consider direct emissions from the harvesting of biofuel crops and indirect emissions from land use changes.

Furthermore, the use of fertilizers causes eutrophication of water bodies and acidification of rainwater, while decomposition of fertilizer generates nitrogen oxides that have an impact on ozone layer depletion. It has been pointed out that these impacts are insufficiently covered by LCAs and require future study (UNEP 2009).

In addition to GHG emissions, land use change can have a potentially major negative impact on biodiversity by changing living creatures' habitats, except where wasteland is used to cultivate energy crops. Jatropha has drawn interest as a raw material crop for biofuel but is also identified as being a potentially invasive species that could disrupt ecosystems.

It is also necessary to consider the nutrient contamination of water bodies as a result of intensive agriculture. There is the additional concern that irrigation and other practices involved in harvesting biofuel crops will also increase consumption of agricultural water and, combined with extreme weather events (flooding and droughts) caused by climate change, could create issues for water resource management (UNEP 2009; World Bank a 2010).

The use and combustion of biofuels reduce localized emissions of some air pollutants such as particulate matter (PM10), volatile organic compounds (VOCs), sulfur oxide, and carbon monoxide but are also reported to increase nitrogen oxide (NOx) and aldehyde emissions. Biodiesels are reported to increase NOx emissions but reduce PM and VOC emissions compared with low-sulfur diesel oil. In many cases, liquefied petroleum gas (LPG) and compressed natural gas (CNG) achieve greater reductions (World Bank a 2010; Arai 2009).

Land use change from biofuel use falls into two categories: impacts that are directly caused by harvesting of biofuel crops and indirect impacts due to changes in the harvesting of other crops which are induced by expanding of biofuel crops harvesting. Both lead to the conversion of land that is needed for forestry or for agriculture to increase food production. This could constrain resources even further. While the predicted impact depends on the target that is set for biofuel use, there are studies indicating that there will not be a major increase in land conversion because of large food demand from China and India and relatively low production of biomass, although land conversion has increased in Africa and Central and South America and to a lesser extent in the USA and Australia (World Bank a 2010). Furthermore, it is thought that natural forests and grassland that are not being used for forestry will be the main target for conversion into biofuel cropping (World Bank a 2010).

It is reported that between 475 and 580 million hectares (Mha) of abandoned agricultural land could be brought back into agricultural production, but not all of these can be returned to productivity, for reasons such as water and nutrient shortages. Furthermore, some countries such as India and China prohibit the conversion of planted forests to return them to agricultural production (World Bank a 2010).

13.1.1.2 Food Security and Biofuels

Manufacturing and use of biofuels are closely connected with food security, namely, the stable supply of food at affordable prices, which is an issue for developing countries in particular.

The Food and Agriculture Organization of the United Nations (FAO) estimates that 925 million people were undernourished in 2010 (FAO 2010) and maintains that food production must increase in order to accommodate future population growth. It is estimated that crop yields per unit area will continue to increase at the recent historic average of 1.5% annually to meet increased demand from population growth, but meeting increased demand for feeds to accompany increased meat consumption will not be possible. This indicates that it will also be difficult to increase the volume of biofuel crops to cope with increased demand for biofuels. While there is a bare minimum expansion of agricultural land necessary for increased food production in response to population growth, there are many estimates regarding the extent of land use changes that will result from changes in food demand and the cropland expansion that will result from increased biofuel use. These estimates vary widely due to differences in underlying assumptions and estimation methodologies (World Bank a 2010). For example, the Gallagher Report (RFA 2008) estimates that between 144 and 334 Mha of additional land will be needed by 2020, equivalent to between 10% and 24% of all cropland in use in 2008.

Food purchasing costs account for a large portion of household expenditures for low-income households, and food security becomes an issue if biofuels cause food prices to rise. In fact, biofuels were one of the factors blamed for the global food crisis of 2008. Other factors blamed included increased demand for grains and meat in emerging countries such as China and India, climate conditions including a drought in Australia, the sharp decline of the US dollar, implementation of export restrictions in food-exporting countries in order to combat domestic food price inflation, and speculative trends in international markets.

Recent reports indicate that the impact of biofuels on food prices is relatively small, both cumulatively and on a global scale. For example, it is estimated that increased global production of biofuels in the 2 years ending in June 2008 only accounted for a little over 12% of the rise in the International Monetary Fund's food price index (World Bank a 2010).

However, looking toward the future, prices for corn will increase by 23–72% by 2020 if countries implement their biofuel deployment plans. The World Energy Outlook 2008 by the International Energy Agency (IEA) projects a scenario in which food prices are calculated to rise by 10% by 2020 if biofuel deployment is maintained at 2008 levels (World Bank a 2010).

The impact of biofuel use on the food supply and food prices must be carefully considered. The joint statement issued by the FAO's World Summit on Food Security (FAO 2009) and the leaders statement of the G8 Hokkaido Toyako Summit both call for a balance between policies promoting the sustainable production and use of biofuels and food security.

13.1.1.3 Rural Development, Agricultural and Industrial Policy, and Biofuels

The FAO points out that new agricultural investment to accompany biofuel use has the potential to create new markets and employment in agriculture, which has struggled with sharply dropping food prices over the last few decades (FAO 2008). According to the FAO, an appropriate increase in biofuel production in rural areas will improve infrastructure development and enhance access to markets. Furthermore, it will help to modernize agriculture and rural economies, improve access to modern energy, and improve indoor air pollution through the use of biofuels that are less polluting to the environment. If good practices such as no-tillage cropping and direct seeding can be employed to harvest biofuel crops, negative impacts, including carbon dioxide (CO_2) emissions and consumption of water resources, can be lessened. In addition, development of local production systems that combine food and energy crops correctly can reduce waste and raise the overall production efficiency.

Notably, the Cabinet of the Japanese Government approved a Biomass Nippon Strategy in December 2002, further revising it in March 2006. The strategy outlines concrete initiatives and a plan of action for encouraging the use of biomass, including biofuels, with the additional perspective of revitalizing agricultural, forestry, and fishing communities. Under the strategy, the Japanese government is implementing policies for significant deployment of biofuels (Government of Japan 2006).

13.1.1.4 Energy Security and Biofuels

The UNEP reports that global production of ethanol for transport fuel increased from 17 billion liters to 52 billion liters between 2000 and 2007, while biodiesel increased from 1 billion liters to 11 billion liters in the same period. Biofuels altogether provided 1.8% of the world's transport fuel, with ethanol providing 5.46% of global gasoline use, while biodiesel provided just 1.5% of global diesel use (UNEP 2009). According to a joint report by the Organisation for Economic Co-operation and Development (OECD) and the FAO, ethanol production is estimated to increase to 159 billion liters by 2019 with biodiesel production increasing to 41 billion liters, which, even considered together, will not account for a significant share of the overall consumption of transport fuel (OECD-FAO 2010). As a result, while the role of biofuels from an energy security perspective varies by country depending on national circumstances, biofuels do not play a major role from a global perspective. However, the situation could change due to uncertain factors including future policies to promote biofuels, economic circumstances such as crude oil prices, environmental standards, developments in second-generation biofuel technologies, and competition between food production and biofuel production from agricultural resources.

13.1.1.5 Trade, North-South Issues, and Biofuels

Global international trade in biofuels only accounts for around one-tenth of all biofuel production, but global trade in ethanol fuels is estimated to have tripled from less than 1 billion liters in 2000 to around 3 billion liters in 2007. The USA is the world's largest ethanol-importing nation, with Brazil the largest exporting nation. More than 10% of all biodiesel production in 2007 was traded internationally, with Indonesia and Malaysia being the major biodiesel exporting nations (World Bank a 2010).

The USA and EU have targeted domestic biofuel production in consideration of their respective domestic biofuel use targets, and no countries other than Brazil currently have the production capacity to become major biofuel-exporting countries. South and Central America and Africa have gaps between their biofuel production potential and actual production and as such have the potential to become exporting countries in the future. In India, trade opportunities are restricted by high tariffs. Although OECD countries have low tariffs for biofuel imports, these countries spend heavily on subsidies to protect their domestic agricultural industries. Furthermore, it has been pointed out that regulations such as the EU's sustainability criteria serve as barriers to trade (World Bank a 2010).

Global trade in biofuels is expected to increase in the future, due to increased demand coming from targets for biofuel use in developed countries, and the potential developing countries have to increase supply through agricultural development. Biofuel imports will also be critical to countries such as Japan that are unable to meet their domestic targets for biofuel use through domestic production alone. The OECD contends that import tariffs on biofuel ingredient crops that are aimed at protecting domestic production and import tariffs on biofuels actually serve as hidden taxes that raise the cost of using biofuels. Furthermore, opening up these markets will reduce the cost and enhance production efficiency as well as decrease dependency on fossil fuels and reduce impacts on the environment (OECD 2008). While certification schemes for biofuels bring about product differentiation based on how biofuels are manufactured and their impacts as determined by life cycle analysis, there is continuing debate about these schemes' relevance to World Trade Organization (WTO) rules when they are used to restrict trade (UNCTAD 2008).

As to the treatment of biofuels in environmental conventions, the Convention on Biological Diversity specifies that production and use of biofuels should be sustainable from a biodiversity perspective and in particular should minimize negative impacts on the lives of indigenous and local communities. The Conference of the Parties (COP) to the Convention on Biological Diversity has issued a decision urging national governments to apply a precautionary approach to the introduction of modified organisms for the production of biofuels, in accordance with the Preamble to the Convention and the Cartagena Protocol (CBD/COP 10 Decision X/37 2010). The implementation framework for the United Nations Framework Convention on Climate Change puts into practice the Reducing Emissions from Deforestation and Forest Degradation (REDD+) program. REDD+ issues funds and credits as economic incentives for reducing CO_2 emissions through clean development mechanism (CDM) projects or efforts by developing countries to restrict deforestation and forest degradation, in the interest of having forests as important carbon stores but also as future sources of material for biofuels. These measures form a response based on the principles prescribed by the Convention on Biological Diversity and Framework Convention on Climate Change, which place common but differentiated responsibilities among developed countries and developing countries, and are important for strengthening international systems to support the sustainable development and use of biofuels.

In any case, it has become essential to respond to north-south issues that accompany financial and technical assistance measures for developing countries and to establish appropriate trade rules in order to use biofuels sustainably at the global level.

13.1.2 Current Study into Sustainable Deployment Strategies

Amid such issues, the United Nations and other organizations are at the center of various international trends such as policy proposals aimed at promoting the sustainable use of biofuels at the global level. Several of these initiatives including UNEP's proposal of measures for the sustainable production of biofuels with an emphasis on environmental aspects as a UN initiative and a study by the FAO done principally from the perspectives of food security and agricultural promotion are introduced briefly in this section. We also describe UN-Energy's principles on sustainable biofuels from the perspective of renewable energy use, and biofuel initiatives by the Global Bioenergy Partnership (GBEP), which was launched by the G8 as a more comprehensive and concrete effort with the involvement of major stakeholder countries and organizations. Lastly, OECD policy proposals that cover economic aspects from the perspective of developed countries are presented.

UNEP released the *Assessing Biofuels* report (UNEP 2009), which was prepared by an international panel on sustainable resource management. The report points out that countries' current biofuel policies do not have adequate scientific bases for their estimates of GHG reductions and that biofuels currently offer only a slight overall potential for GHG reduction, while the costs so far, as identified by the OECD (OECD 2008), are extremely high. It also states that the sustainable production of biofuels is achievable if a strategy is implemented to enhance resource productivity and identifies four measures to enhance resource productivity: (1) the use of obligations for biofuel use and biofuel targets and standards, encompassing the development of resource management programs by country and region, and development and implementation of standards and certifications for biofuel production; (2) promotion of sustainable land use for biomass production, encompassing measures such as the study of comprehensive guidelines for land use management; (3) more efficient use of biomass including the use of residues and waste and cascading use of biomass; and (4) increased energy and material productivity in transport, industry, and households as the basis for advancing a low-carbon, recycle-oriented society.

The FAO pursues efforts with a particular emphasis on five policy principles and areas, focusing in particular on the relationship between food security and biofuels. These are the following: (1) protecting the impoverished and their food security, including the stable supply of food at appropriate prices to developing countries that import food; (2) using opportunities for agricultural and rural development, including financing and technical support to small-scale farmers; (3) securing the environmental sustainability of efforts to create climate change mitigation benefits, through the study of sustainability standards and certification systems and the deployment of biofuels; (4) reviewing existing biofuel policies encompassing the review of trade barriers for biofuels, the shift to second-generation biofuels, policy consistency such as for carbon taxes and emissions trading, and a less rapid shift to biofuels; and (5) strengthening international systems for supporting sustainable biofuel development, including mechanisms for achieving environmental targets that use sustainability criteria agreed on in an international forum. Currently, through the Bioenergy and Food Security Project, FAO is studying principles, criteria, and guidelines for biofuel use, with an emphasis on reducing trade barriers, reaching agreement on international sustainability criteria (standards), and capacity development in developing countries from an international perspective.

At the 2002 World Summit on Sustainable Development (WSSD), 20 UN organizations formed the UN-Energy program with the aim of securing consistency in UN organizations' multidisciplinary response. Regarding bioenergy, which includes biofuels, UN-Energy has identified key areas for implementing the sustainable use of biofuels and energy at the international level. Specifically, UN-Energy points to the need for monitoring and assessment of the impact of bioenergy development on agriculture, industry, health, environment, and trade and for sustainable cropping and use of energy crops conforming to the mechanisms of environment conventions such as the Framework Convention on Climate Change. UN-Energy further points to the necessity of technology development for bioenergy use and the establishment of standards and certification systems in a way that does not obstruct trade, as well as the need for technology transfer and development. UN-Energy has also identified items for national policymakers to study in deploying policies for biofuel use (UN-Energy 2007).

At the G8 Gleneagles Summit 2005, the G8 leaders agreed on the Gleneagles Plan of Action for climate change, clean energy, and sustainable development and made the decision to launch the Global Bioenergy Partnership (GBEP) to support wider, more cost-effective biomass and biofuels deployment, particularly in developing countries where the use of biomass is prevalent. Participants in the GBEP include the G8 countries, developing countries such as Brazil and China, and international organizations such as the FAO and IEA. The GBEP conducts study into efficient policies for supplying rules and tools for promoting sustainable bioenergy (biomass and biofuels). Specifically, these efforts are in the following three areas: developing voluntary standards and indicators that are practical and have a scientific basis to promote the sustainable development of bioenergy, testing a common methodological framework for measurement of GHG emission reduction from bioenergy use; and awareness raising to promote information exchange regarding bioenergy.

Every year, the GBEP reports the results of its studies to the G8 and G20. In particular, the GBEP is aiming to identify criteria and indicators that are consistent with multilateral trade agreements but intended to be used at the domestic level. In May 2011, the GBEP reached agreement on a list of indicators to report to the G8 Summit in 2011. The GBEP indicators comprise eight items each in the three areas of environmental, social, and economic and energy indicators. Environmental indicators include GHG emissions from a life cycle perspective, percentage of water resources used for harvesting and manufacturing, biodiversity, and changes in land use. Social indicators include legal instruments for the distribution and ownership of land for new bioenergy production, the price and supply of food, and changes in income, job creation, and time spent collecting biomass by women and children. Economic indicators include productivity, net energy balance, and diversity of energy supply sources as affected by the supply of biofuels (GBEP 2011).

The OECD has issued policy proposals for the assessment and deployment of biofuels, principally from an economic and trade perspective (OECD 2008). These policy proposals hold that there is no one best common policy for biofuel use and that it is necessary for each country to use an appropriate combination of policies that match their policy priorities and social and economic conditions. Furthermore, in order to conserve energy, it is necessary to move toward lower energy consumption and enhancing energy efficiency rather than substituting with biofuels. The growth of the biofuels sector also raises food prices and reduces food security in the medium term for the most vulnerable people in developing countries. As such, the OECD proposes establishing ambitious minimum standards for GHG reduction by biofuels while avoiding the harvesting of biofuel crops in environmentally sensitive lands such as wetlands. It also proposes opening international markets to bring about more efficient production of biofuels. These measures would reduce unintended secondary effects and lead to enhanced employment and income opportunities in developing countries through responsible trade.

In addition, the OECD contends that further study of the environmental risks from land use changes is needed, encompassing high-efficiency production of biofuels in tropical and subtropical regions as well as indirect land use changes.

13.1.3 Tools for Sustainable Use Strategies

13.1.3.1 Policy Tools for Sustainable Use Strategies

An overview of current challenges and policy recommendations relating to the sustainable use of biofuels was introduced in previous sections. Clearly, it is necessary to approach the sustainable use of biofuels in a way that takes into account environmental conservation issues, food security, community development, energy security, trade, and the "north-south divide." We show that the Food and Agriculture Organization (FAO), United Nations Environment Programme (UNEP), and other organizations are issuing policy recommendations that meet these needs. Here, we focus on concrete policy methods by extracting strategy tools from such recommendations. These proposed policy methods can be broadly divided into eight categories outlined below. These are closely related to each other. For example, where standards, indicators, and certification systems are applied, sustainable land use is promoted by the standards and methods if they are set appropriately. From an international perspective, some of the more important strategies and initiatives are development of standards and indicators, the application of certification systems, international market liberalization, and technology transfer to developing countries.

A. Development of Standards and Indicators and the Application of Certification Systems for the Sustainable Production and Use of Biofuels

The setting of standards for CO_2 emission and land use conversion, based on LCA (life cycle assessment), and for strict legal compliance and community consultation makes it possible to evaluate and manage not only the environmental impacts of biofuel use and production but also impacts on a broad range of other areas, including community development and trade, by regulating social and economic impacts and sustainability. In addition, combining standards and indicators with a certification system guarantees the effectiveness of those standards. In the EU, for example, the introduction of a voluntary certification system is being recommended.

B. Promotion of Sustainable Land Use for Biofuel Crop Production (Including Land Use Management Planning and Increasing Yields)

To increase the production of food or biofuel energy crops, it is essential to increase agricultural crop yields. This requires expanding farmed areas, which means developing into precious natural ecosystems. Expanding the cultivation of biofuel energy crops has both direct and indirect impacts. For this reason, it is important to try and improve yields per unit area using methods that are both environmentally and people friendly. And with the aim of sustainable land utilization, land use management plans and guidelines must be developed that take into account agriculture, forestry, mining, and other industries at regional, country, and international levels.

C. Reviewing Support Systems for Efficient Production and Distribution, Liberalization of International Markets, and Lowering of Trade Barriers

In addition to providing longstanding protection of domestic agriculture, sometimes subsidies are provided and import duties are levied on raw biofuel materials and products to hasten the expansion of biofuel utilization. By reviewing such financial measures, it is possible to promote more efficient and inexpensive production, reducing environmental impacts and dependence on fossil fuels (OECD 2008). Revising these measures can also serve to stimulate increased production at a sustainable level and to steer biofuel production to the most efficient regions and countries. Fair trade also leads to improved opportunities for employment and income in developing countries.

D. Transfer of Technology for Cultivating Biofuel Crops to Developing Countries

The transfer of technology to developing countries for increasing biofuel crop yields and improving of fermenting raw materials and recovery rates tends to reduce environmental impacts due to improved biofuel crop cultivation and biofuel production.

E. Efficient Use of Biomass and Biofuels

It is important to investigate the connection between the use of biofuels with biomass use, taking into account the second-generation biofuels that are expected to be widely diffused in the near future, as well as their use in generators, or so-called stationary facilities. As pointed out by the UNEP (UNEP 2009), the use of waste matter and production residue, and "cascading" (first using biomass as a raw material for production and then recovering the energy of the waste matter generated by production), increases the potential for reducing CO_2 emissions from the biomass. In addition, it is generally reported that the use of biomass is more energy efficient when used in stationary facilities rather than for transport. These methods indirectly reduce the demand for biomass fuels and thereby enable their use to be limited to a level that can be met by sustainable production.

F. Improving Energy Efficiency for Transport, Industrial, and Household Uses

Improving energy efficiency makes it possible to reduce overall demand for liquid fuels. As a result, the demand for biofuels, as fossil fuel substitutes, also decreases, improving their sustainability.

G. Promotion of Surveys and Research Assessing the Value of Ecosystem Services and Developing Second-Generation Biofuel Production Technology

To address the impact of expanding biofuel usage on ecosystem services and develop measures to address such impacts, one of the options proposed by the United Nations University for ensuring sustainable use of biofuels involves paying for ecosystem services. Evaluation methods, however, need to be investigated in further research. Second-generation and more advanced biofuels, made by cellulose decomposition, are not yet cost competitive, so it is necessary to promote further study and research to address this issue.

The UN-Energy program, UNEP, and United Nations University Institute of Advanced Studies (UNU-IAS) have all recommended sustainability standards, indicators, and certification systems as effective tools in strategies to promote sustainable biofuel utilization at an international level. The Global Bioenergy Partnership (GBEP) is also examining their introduction. Because such measures have already been partially implemented and proven to some extent in international consensus and practice, in ISO 14001 and other fields like forestry, they are now the center of attention and are being studied in light of the latest trends.

13.1.3.2 Outline of Standards, Indicators, and Certification Systems for Sustainable Biofuels

Like the certification system of the ISO 14001: environmental management standard, standards, indicators, and certification systems are designed to promote measures that counter adverse environmental impacts by defining the standards and indicators necessary to achieve specified targets of environmental conservation. Products and enterprises that meet those requirements can then be awarded certification. Standards define the concept of sustainability, while indicators serve as quantitative, or in some cases qualitative, criteria for measuring and assessing compatibility with the standard. Standards and indicators can be used independently to define policy goals or as part of a certification system to define specific certification criteria that differ from existing standards and indicators.

A certification system that covers all stages from production through processing and distribution is referred to as "chain of custody certification" or "COC certification." One example of such a chain is the processing of ethanol or vegetable oil from a certified farm, followed by fermentation/extraction and proper delivery to consumers. Since biofuels are liquids, the risk exists that they may become mixed up or confused in the process of distribution, either with other certified fuels or with an uncertified fuel. For this reason, measures such as those below have been introduced for tracking biofuels through the certified product supply chain.

- Identity Preserved (IP): In this system certified biofuels are separated from the plantation and tracked until reaching the user.
- Segregation (SE): In this system mixing of certified biofuels (batch mixing) is recognized and tracked.
- Mass Balance (MB): In this system, if a certified biofuel is intended to be mixed with another biofuel, the quantity of that particular certified biofuel is defined as a proportion of the total certified biofuel. A manager monitors the mixing proportions, which depend on the fuel's final use.
- Credit Trading (CT): This system does not involve any tracking, tracing, or monitoring of the biofuel itself. Instead, it enables cultivators and users to trade volume credits online (known as "Book and Claim (BC) in the case of the Roundtable on Sustainable Palm Oil (RSPO), Bangun 2011) under the supervision of managers, as done with green electric power, for example.

These kinds of reference standards and indicators and certification systems are already utilized in the fields of forestry and marine products. Good examples are the certification system of the Forest Stewardship Council (FSC) and the system implemented by the Marine Stewardship Council (MSC). (See the FSC and MSC websites.)

The features of standards and indicators and certification systems based on them are outlined below (UNCTAD 2008; UNEP 2009; Scarlat 2011).

A. Diverse Principles, Standards, and Indicators Can Be Defined to Address the Various Environmental, Social, and Economic Impacts of Biofuels

In addition to dealing with principles concerning the reduction of greenhouse gas emissions, taking into account product life cycle and factors such as land use change, many certification systems currently being developed also deal with other important issues. These may include environmental considerations, such as biodiversity conservation and land use change; social considerations, such as food security and social well-being; and economic considerations, such as productivity. In addition, these systems can also incorporate local viewpoints by recognizing the particular interpretations of individual countries, as in the case of the RSPO. They can also establish an international system, like the sustainability standards for bioenergy now under consideration by the International Organization for Standardization (ISO), and incorporate global-scale, unified standards and indicators in alignment with international agreements such as the UN Framework Convention on Climate Change and WTO treaties.

B. Certification Is Not Mandatory

Choosing to certify a product makes it possible to differentiate it, by providing buyers with information about sustainability relating to the product. In particular, COC certification can assure buyers that they are definitely getting a product that meets specified standards and indicators. In this way, producers can add value to products, thereby benefitting themselves. Certification can even have a substantial impact on market share.

C. In Combination with Other Policy Measures, a Certification System Can Drive Initiatives That Mitigate the Impacts of Biofuels

Certification can be linked to tax deductions and other incentives, and as in the case of the EU and USA, it can serve as a precondition for measuring product consumption to reach the national goals.

On the other hand, there are limitations and problems with standards, indicators, and certification systems, such as the following:

A. Scope and Effectiveness of Standards and Indicators

In setting standards and indicators, it is possible that some problems will be ignored. For example, when making evaluations based on limited LCA criteria, it can be difficult to quantify social standards. Also, since the evaluation of macro effects, such as the impacts of biofuel production on food and land prices and forest depletion, depend on the adopted methods of evaluation, there is a risk that subjective points of view can creep into the analysis. Although it is necessary both that criteria are comprehensive and that a standard is technically and administratively practicable, these requirements cannot necessarily be satisfied. Note that there is potential for confusion when trying to make categorizations, i.e., certification is needed for a crop used for biofuels but not for the crop consumed as foods. In policy matters as well, this can lead to two kinds of standards, resulting indirectly in an increase in land use conversion.

Addressing indirect land use change would require implementing certification over the whole planet, in order to get a complete picture of the system. However, since this is totally unrealistic, standards, indicators, and certification systems cannot be very effective in dealing with indirect land use change. Devices such as the iLUC factor (indirect land use change factor) for specific products, i.e., the evaluation based on a global-scale life cycle for each consuming country/region and product/production process, are necessary, but data usability is still a problem.

B. Coexistence of Multiple Certification Systems

There are many certification systems now in existence, but the fact that they are all different tends to reduce their effectiveness and reliability. Multiple systems also tend to result in market fragmentation and reduced transparency. For this reason, only international certification schemes can achieve environmental goals. However, it must be noted that certified products only account for a small part of the market. It is also important that developing countries, farmers' associations, and local NGOs are properly represented and are able to contribute to the international processes of creating standards, indicators, and certification systems and to reaching agreements between countries. Unfortunately, this is not necessarily the case at present.

As opposed to creating new standards for biofuel crop production, especially for biofuels, the "meta-standard approach" makes use of existing standards for sustainable agriculture and forestry. The use of existing standards offers numerous benefits: assured reliability, easier acceptability to buyers, quick implementation, greater cost effectiveness, less confusion between different standards, and promotion of convergence (Committee on the Sustainability Criteria regarding Introduction of Biofuels 2009).

C. Cost of Certification

Firstly, the cost of certification to producers consists of the cost of conforming to standards and indicators plus the cost of acquiring certification. Although these costs depend on the number, stringency, and comprehensiveness of the standards and indicators defined by the certification system, production costs are likely to increase substantially. In addition, the cost of certification is likely to be more burdensome in developing countries than in developed countries and to small-scale producers than large-scale producers.

D. Connections to Trade

In the years ahead, the role of international trade is expected to become more important, which should lead to more effective utilization of biofuels. However, to assess the connection of biofuels with WTO agreements, which set the general rules of international trade, it should be noted that the present level of biofuel production and international trade is relatively small and that certification systems for biofuels are quite new. In view of this, there is not yet any established view how biofuel certification systems can be dealt with under the current international trading rules (UNCTAD 2008).

More specifically, there is no clear consensus about whether certification systems developed by NGOs or other private organizations fall under WTO rules or whether they should be regarded merely as marketing strategies. Although certification serves to differentiate products based on their methods of production and their LCA-determined impacts, any differentiation based on process and production methods (PPM) may be in violation of WTO agreements. Also, there is still some debate about whether certified biofuels can be justified as exceptions to the rules (under Article 20 of GATT) as measures necessary to protect human, animal, and plant life and health or measures to conserve limited natural resources. And doubts remain about the differentiation of products on the basis that they meet a broad range of objectives such as workers' rights and food security or on the basis of their production process. Note that no study appears to have been done on certification systems in relation to government support schemes or subsidies in light of international trade agreements (UNCTAD 2008).

For example, under the EU Renewable Energy Directive, the standard relating to biodiesel fuels stipulates that they should not be cultivated on converted peat lands and that they should reduce CO_2 emissions by at least 50% relative to conventional fuels on a life cycle basis. The directive recommends the use of a voluntary certification system to prove these requirements. However, the Indonesian government and US soybean producers have expressed fears that this standard violates WTO rules relating to PPM (Jakarta Globe 2010, GlobalSubsidies 2011 HP).

At the same time, some countries give "favored nation" treatment to particular international trading partners. For example, under a preferential import system for ethanol fuels in the USA, ethanol imports from Caribbean Basin Economic Recovery Act (CBERA) countries are exempt from import duties, although there is an upper limit on import volume. In 2006 these countries accounted for 25% of all imports of ethanol for fuel use, but until 2003 they accounted for 100% of the total import volume (Uchida 2007).

As the OECD has pointed out, eliminating this kind of trade barrier through an international consensus on WTO rules relating to certification systems can lead to a liberalization of international markets that gives rise to more efficient global biofuel production.

13.1.3.3 Current State of Standards, Indicators, and Certification for Sustainable Biofuels

According to the UNEP, there are at least 29 sustainable biomass or biofuel-related initiatives to establish standards, indicators, or certification systems presently being conducted by various national governments, NGOs, worldwide organizations, and other bodies (UNEP 2009). As of January 2011, the FAO's Bioenergy and Food Security Criteria and Indicators Project (BEFSCI) was dealing with 17 initiatives, reviewing outlines of regulatory frameworks for the EU and other regions (5 cases), voluntary standard and certification schemes (10 cases), and scorecards (2 cases), with most of these cases being related to biofuels.

There are currently many activities in progress all over the world, but here, in accordance with the FAO classification, we will look at an outline of the main initiatives relating to voluntary standards, indicators, certification systems, and regulatory frameworks (See Tables 13.1 and 13.2).

A. Voluntary Standards, Indicators, and Certification Systems

The Global Bioenergy Partnership (GBEP) and the Roundtable on Sustainable Biofuels (RSB) are aiming to develop nonbinding, voluntary standards and indicators for sustainable biofuel use with a scientific foundation that is available worldwide. Criteria (standards) are defined as categories like sustainability elements, capacities, or processes that are employed for evaluating the environmental, economic, and social performance of bioenergy production and utilization. Indicators, on the other hand, are measurable outcomes based on the criteria. They are considered the means for measuring or describing the various perspectives of the criteria. Of the perspectives represented by the sustainability standards, the environmental, social, and economic perspectives are indispensable. Thus, indicators are required to enable appropriate evaluation of these three kinds of issues.

GBEP has recently agreed on 24 indicators in the three fields but found that the initial indicators they proposed to gage indirect impacts on land utilization due to

Table 13.1 Classification of standards and indicators on sustainable biofuels

Categories	Example of activities	Outline
1. Voluntary standards, indicators, and certification systems	Global Bioenergy Partnership (GBEP)	Development of voluntary standards and indicators by G8 initiative
1.1. International and regional activities	ISO, CEN	Voluntary standard, indicators, and certification systems developed by businesses and other stakeholders worldwide or European wide
A. Governments and international organizations		
B. NGO and others	RSB	Voluntary system of standards, indicators, and certification of stakeholders lead by EPFL
1.2. Activities on each biofuel energy crop	RSPO (oil palm), RTSP (soybean), BONSUCRO (sugarcane)	Voluntary systems of standards, indicators, and certification systems for each crop developed and agreed upon by stakeholders including producers
2. Regulatory framework in a region or country	EU, Germany, the Netherlands, Brazil, USA	Standards used by governments in order to comply with "fuel mixing targets" and/or "consumption targets" of biofuels
3. Score cards	IDB	Systems to improve the performance of a project by scoring systems through assessing environmental impacts, socioeconomic impacts, and impacts on food security of a biofuel project
	WB/WWF	

CEN the European Committee for Standardization, *IDB* Inter-American Development Bank, *WB/WWF* World Bank/World Wildlife Foundation

cultivation of bioenergy plants and indirect impacts relating to the price of agricultural products require further study. Also, indicators do not serve to express the direction and threshold values of measures and standards but to express the state of progress toward sustainable development in individual countries (GBEP 2011).

RSB is an organization led by the École Polytechnique Fédérale de Lausanne, consisting of more than 720 diverse stakeholder organizations, including biofuel users, producers, policymakers, companies, and financial institutions. The body is currently engaged in an initiative aimed at creating tools to help these stakeholders make judgments about sustainability. It recently released version 2 of its RSB Guidelines, which outline principles and standards for global-scale sustainable biofuel production, and in March 2011 launched a certification system connected to these guidelines. This system conforms with regulations on biofuel requirements set by the government of Germany, which aims at expanding the use of biodiesel and will become an important importer in the future, as well as biofuel regulations based on the EU directive.

A survey of initiatives relating to specific biofuel crops reveals that there are current international initiatives by bodies connected with the oil palm, soybean, and sugarcane industries.

Table 13.2 Overview of standards and indicators of biofuels

Name	GBEP	RSB Ver.2	RSPO	RTRS	Bonsucro	EU	USA	Japan (proposal)
Features and Scope	Voluntary					Compulsory		
	Intergovernmental	Multi-stake holder				Regional, country specific		
	International					Regional	Country specific	
	Crops in general		Palm oil specific	Soybean specific	Sugarcane specific	Crops in general		
Specific certification system established	–	X	X	X	X	X([a])	–	–
Environment	X	X	X	X	X	X	X	X
Reduction of GHGs (GHGs emissions)								
Land use change([b])	X	X	X	X	X	X	X	X
Carbon stocks								
Others (soil, water, air, biodiversity, waste management, ecosystem conservation, natural resource and utilization and energy saving, env. impact assessment, agricultural chemicals and integrated pest management, etc.)	X	X	X	X	X	X	X	–
Economic (productivity, promotion of tech. Rural development, good practice, long-term economic viability, etc.)	X	X	X	X	X	–	–	–
Social	X	X	–	X	–	X	–	X
Food security, poverty/social development								
Human and labor rights, health (occupational)	X	X	X	X	X	X	–	–

(continued)

Table 13.2 (continued)

Name	GBEP	RSB Ver.2	RSPO	RTRS	Bonsucro	EU	USA	Japan (proposal)
Land rights	X	X	X	X	X	X	–	–
Good governance (stakeholder consultation), legality (compliance)	–	X	X	X	X	X	–	X
Other	–	X	X	X	X	X	–	–
Planning, monitoring of implementation, transparency, COC requirement, etc.)								

X: Items included
GBEP: Global Bioenergy Partnership 2011
EU: Directive 2009/28/EC, the promotion of the use of energy from renewable sources
USA: Renewable Fuel Standard 2010, EPA must estimate impacts of RFS other than CO2 reduction
JP: Report of the Committee on the Sustainability Criteria regarding Introduction of Biofuels, Japan (in Japanese) 2009
RSB: Roundtable on Sustainable Biofuels, Version 2.0 2010
RSPO: Roundtable on Sustainable Biofuels, 2007
RTRS: Round Table on Responsible Soy Association, Version 1.0 2010
Bonsucro, Better Sugar Cane Initiative, 2010
[a]Specific certification systems have not been established, but voluntary certification of other systems can be used
[b]Impacts of indirect land use change are not included in the current

a. *Oil Palm*

Palm oil, obtained from the fruits of oil palms, is one of the most abundant vegetable oils in the world, with global production estimated at approximately 46 million tons (2010 estimate, Yushi (Oils and Fats) 2011). In addition to its use in food products such as cooking oil, margarine, and shortening, it is used as a raw material for soap and increasingly for the production of biodiesel fuel. The main producers of palm oil are Indonesia and Malaysia. Although the proportion of their palm oil output that is used as a raw material for biofuels is relatively small, demand for food products is growing, and the conversion of forests and peat lands to plantations has become a problem. In view of this, in 2003 the Roundtable on Sustainable Palm Oil (RSPO) was formed to promote sustainable methods of oil palm cultivation and palm oil production. RSPO members include oil palm growers, palm oil processors and traders, consumer good manufacturers, retailers, banks and investors, and NGOs. Currently, there are more than 400 members and 110 supporting members.

RSPO has set forth eight principles and 39 criteria to promote sustainable production and consumption. For example, its environmental standards and indicators stipulate that after 2005 new plantations must not be converted from virgin forest or areas of high conservation value and that levels of pollution and waste products must be reduced. To prove that these standards and indicators are strictly observed, a certification system was introduced for plantations and extraction plants, and a COC certification system was introduced for the supply chain. For regular certification, the certifying body visits the plantation or plant to conduct an auditing, during which they examine documentation and the site/facilities and interview relevant parties. A summary of the auditing report is then posted on the Internet. For COC certification, in addition to an identity preserved (IP) system that allows the use of palm oil produced by certified facilities, segregation (SG) and mass balance (MB) systems were also adopted. Book and Claim (BC) was also introduced for credit trading. Also, to promote awareness among consumers, RSPO developed a trademark, which is expected to come into use in 2011. As of January 2011, there were 81 certified processing plants and a total of approximately 760,000 ha of certified plantations in Malaysia and Indonesia, and 3.8 million tons of palm oil has so far been certified (Bangan 2011). Nonetheless, RSPO still faces significant problems. Standards and indicators relating to greenhouse gases are still only under consideration, and the demand for certified palm oil is low relative to the supply.

b. *Soybeans*

Soybeans account for approximately 29% of the total worldwide production of vegetable oil, second only to palm oil. Like the palm oil industry, in 2006 the Roundtable on Responsible Soy (RTRS) was founded, with the aim of promoting the responsible production and use of soybeans, through the participation of stakeholders and the application of international standards. The membership of RTRS is made up of growers, the soybean oil industry, traders, financial institutions, and NGOs, while individuals and governments can participate as observers. As of 2009 the total membership was approximately 110 organizations, of which half were producers. In May 2009, principles and standards for field testing were approved by the general assembly, and a pilot project was launched. Then in 2010 standards developed from five principles were adopted at the general assembly (RTRS 2010). Work is also in progress on preparing the implementation of a certification system, under a similar framework to that of the palm oil industry, and a certification trading platform corresponding to palm oil's book and trade feature was expected to go into effect in April 2011. Note that it has been agreed that soybean oil certified by the RTRS in cooperation with the EU can be considered as a biofuel that complies with the requirements of the EU's Renewable Energy Directive (RED), provided that it satisfies specific requirements.

c. *Sugarcane*

Bonsucro is an international nonprofit organization that aims at social, environmental, and economic sustainability of sugarcane-related activities by promoting the development and use of global standards. The membership is similar to those of the other crops mentioned, e.g., sugar producers, NGOs, and other stakeholders. In March 2011, Bonsucro introduced a certification system. It formulated Bonsucro's production standards and developed a set of indicators based on five principles, such as strict compliance with applicable laws. The requirement for certification is 100% compliance with core indicators and 80% compliance with other indicators (Bonsucro 2011). The body has also developed a supply chain certification scheme, making use of MB-based methods. The first supplies of certified sugar are expected to be available around April 2011. Provisions have also been made for compliance with the EU's Renewable Energy Directive (RED) and Fuel Quality Directive by including two corresponding sections in the standards. Compliance with these sections will be considered to represent compliance with the EU directives.

B. Regulatory Frameworks in Countries and Regions

Just as US and EU standards and indicator systems are making a substantial impact on the world, through the international biofuels market, standards and indicators at the national level have become powerful policy instruments.

In USA, life cycle-based standards relating to the reduction of greenhouse gas emissions have been adopted for biofuels used for transportation equipment. In August 2005, the Renewable Fuel Standard (RFS), which mandates the use of biofuels for powering automobiles, was incorporated into the 2005 Energy Policy Law. The standard stipulated the use of a total of 4 billion gallons (approx. 15 million kL) of biofuels by 2006, with a steady increase in subsequent years up to a total of 7.5 billion gallons (approx. 28 million kL) by 2012. Then in December 2007 the Energy Independence and Security Act of 2007 was introduced, which unveiled medium-term policy guidelines on improving energy efficiency and expanding the production of renewable energy. By upgrading the RFS (to RFS-2), the act set even more ambitious requirements for biofuel production: 9 billion gallons per year by 2008, increasing in stages to 36 billion gallons by 2022. It also stipulated that new biofuels derived from raw materials other than corn must account for at least 21 billion gallons of the 36 billion gallon target for 2022. Furthermore, biofuels must account for at least 20% of all fuel for transport by 2020. Under this RFS program, biofuels are classified into four types, with targets set for the quantity of each type used. These biofuels must comply with a standard for cutting greenhouse gas emissions relative to fossil fuels. The US Environmental Protection Agency is also assessing other environmental impacts of RFS-2. For example, the production and use of regular ethanol produced from corn must generate at least 20% less greenhouse gas emissions than regular gasoline, based on LCA, taking into account major indirect impacts such as land use change (Hill 2011).

In the EU, the Renewable Energy Directive (2009/28/EC) requires that by 2020, renewable energy must account for at least 20% of the total final energy consumption in the EU. Targets for biofuel adoption for transport fuel are 5.75% by 2010 and 10% by 2020. The biofuel sustainability standard for this purpose features greenhouse gas reduction targets relative to fossil fuels, based on LCA (min. 35%), environmental impacts (e.g., not permitting raw materials for biofuels from areas of high biodiversity or high carbon storage capacity), and social impacts. As for indirect impacts, a report by the EC concludes that if the share of first-generation biofuels derived from agricultural crops is held to less than 5.6%, and second-generation biofuels are used for the remainder, then biofuels can be very useful in cutting CO_2 emissions, even when additional emissions due to indirect land use change are taken into account (EC 2010).

In addition, in order to verify that biofuels are complying with the above standards, voluntary certification systems are to be used.

Some individual countries, including the UK and Germany, have also formulated their own standards on biofuels in line with EU regulations. In Japan, a report from a workshop on biofuels sustainability ("Towards the formulation of a Japanese standard for biofuel sustainability") issued in 2009 indicated that the effectiveness of greenhouse gas emission reductions based on LCA, land use change, and the stability of supplies of biofuel crops that compete with food for crops are to be studied, as elements of standards and indicators, taking into consideration systems and methods for their operation.

As presented above, standard indicators are divided broadly into five fields: environmental, economic, social, energy, and factors relating to the monitoring of implementation status. The factor to which most attention is paid is greenhouse gas emissions in the environment. For this, quantitative standards (e.g., 35% reduction by the EU and 50% reduction by RSB) are often formulated, with reference to regular fossil fuels, based on LCA. Land use change is also considered in association with greenhouse gases; in the case of GBEP, for example, indicators will be studied further and formulated in the future that account for the indirect impacts of land use change. Other environment-related standards and indicators are generally formulated dealing with soil, water quality and quantity, air, waste, and biodiversity. Standards for specific crops also prescribe the implementation of environmental assessments before development, the integration of pest management, and the use of persistent agricultural chemicals.

Economic standards define productivity assurance, long-term economic viability, and the implementation of best practices. GBEP has a standard related to energy security such as net energy balance, energy diversity, and flexibility of the use of bioenergy, which Japan also emphasizes.

In the field of social standards, GBEP and RSB place importance on worldwide impacts, such as the assurance of food security and community development. At the same time, standards for specific products are concerned with workers' rights and health, land rights, regulatory compliance, and participation in stakeholder evaluations.

Implementation standards are usually concerned with the implementation status monitoring and the assurance of transparency by means such as information disclosure.

Note that EU and US standards and indicators are mainly concerned with CO_2 emissions and associated land use change. As for other standards, the EC and EPA have created separate reports for submission to their national assemblies.

In the case of Bonsucro, it incorporated its COC certification requirements into standards and indicators, but other initiatives set separate regulations for its certification system.

C. Conclusions

In order to improve standards, indicators, and certification systems effectiveness as tools in biofuel utilization strategies on an international scale in the years ahead, it will be important to develop and improve these measures in the directions outlined below, taking into account the current state of utilization and the pros of cons of these tools.

- Unifying existing programs and establishing internationally agreed principles, standards, and certification systems for biofuel sustainability that allow for flexibility for the environmental and socioeconomic diversity of various producing countries and which are quantitative, verifiable, and scientific and formulating systems to develop standard and certification by means of a participatory process in which the stakeholders of various regions are effectively represented
- Establishing standards, indicators, and certification that permit support for small-scale producers, particularly those in developing countries, and providing support for developing countries in improving their capacity to verify compliance
- Developing precise methods for evaluating macro impacts, such as the indirect impacts of using land for biofuels, and practical frameworks that enable rational implementation for highly cost-effective certification
- Investigating systems carefully, to ensure they promote sustainable development through trade, taking into consideration compatibility with WTO rules

13.1.4 Conclusion and Further Prospects

This paper describes the challenges of investigating the global sustainability of biofuels, policy recommendations by UNEP and other bodies, and the current state, features, and issues relating to certification, as an important tool to use in developing sustainable utilization strategies. The results bring many points to light. Firstly, the promotion of rapid, large-scale utilization of biofuels contributes to sharp rises in food prices, and with current technology, biofuel production is not economically viable without subsidies by producing countries, with some exceptions. In addition, in terms of absolute volume, biofuel production will remain relatively low comparing to fossil fuels, and large-scale substitution will be difficult in the short term. Standards, indicators, and certification systems relating to sustainable biofuels are being developed at various levels around the world. If these can be unified on an international scale and if conformity with international trade rules can be maintained, biofuel sustainability has the potential to play substantial roles in, for example, realizing ambitious standards for reduction of CO_2 emissions.

Currently, as part of the broad trend toward a global, sustainable society, various national- and international-level initiatives based on the UN Framework Convention on Climate Change and the Convention on Biological Diversity are being undertaken, and at the UN World Summit on Sustainable Development (UNCSD, RIO+20) to be held in 2012, one of the main agenda items for discussion will be a "green economy." As one element of renewable energy, biofuels are expected to make a significant contribution to the green economy (UNEP 2011). In addition, initiatives directed toward a sustainable planet will be in progress; 2014 will be the tenth and final year of the UN Decade of Education for Sustainable Development, and in 2015 the UN Millennium Development Goals (MDGs) are due to be achieved. At the same time, research is proceeding vigorously on the technological possibilities of manufacturing second-generation and more advanced biofuels and manufacturing biofuels from algae. In Japan, expectations of renewable energy are likely to keep growing in the aftermath of the recent nuclear reactor crisis. In view of all this, it is vital to keep formulating and implementing sustainable biofuel utilization strategies, linking them to local and national-level strategies, and taking into account the survey results reported here.

Reference

Arai (2009) Shinichi Arai, impacts of increasing use of biofuels on water resources and atmospheric environment. J Environ Inf Sc 38(3):50–55. 2009 (in Japanese)

Bangun (2011) Derom Bengun, Sustainable Production on Palm oil by RSPO Standard, Symposium on the Sustainability of Biofuels, February 2011, Tokyo, Japan

BONSCRO (2011) BONSCRO production standards including BONSACRO EU Production Standard http://bonsucro.com/site/wp-content/uploads/2013/03/Bonsucro_Production_Standard_March-2012__c.pdf

Committee on the Sustainability Criteria regarding Introduction of Biofuels, Japan (2009) Report of the Committee on the Sustainability Criteria regarding Introduction of Biofuels, Japan (in Japanese), http://warp.da.ndl.go.jp/info:ndljp/pid/8231957/www.meti.go.jp/press/20100305002/20100305002-1.pdf (バイオ燃料導入に係る持続可能性基準に関する検討会 2009:バイオ燃料導入に係る持続可能性基準等に関する検討会報告書)

EC (2010) Report from the Commission on indirect land-use change related to biofuels and bioliquids (COM (2010) 800 final), European Commission

FAO (2008) The state of food and agriculture 2008, Biofuels: risks and opportunities http://www.fao.org/docrep/011/i0100e/i0100e00.htm

FAO (2009) Declaration of the world summit on food security http://www.fao.org/fileadmin/templates/wsfs/Summit/Docs/Final_Declaration/WSFS09_Declaration.pdf

FAO (2010) The state of food insecurity in the world, http://www.fao.org/docrep/013/i1683e/i1683e.pdf
GBEP (2011) The Global bioenergy partnership, sustainability indicators for bioenergy First Edition, http://www.globalbioenergy.org/fileadmin/user_upload/gbep/docs/Indicators/The_GBEP_Sustainability_Indicators_for_Bioenergy_FINAL.pdf
Hill (2011) Jason Hill, Assessing biofuel sustainability : Lessons from growth of the US industry, Symposium on the Sustainable Biofuels, Feb. 2011, Tokyo
Jakarata Globe (2010) Indonesian biofuel dispute may burn EU, JakarataGlobe 20th June,2010
OECD (2008) Biofuel support policies: an economic assessment http://www.oecd.org/tad/agricultural-trade/biofuelsupportpoliciesaneconomicassessment.htm
OECD-FAO (2010) OECD-FAO agricultural outlook 2010–2011 http://www.oecd.org/site/oecd-faoagriculturaloutlook/48202074.pdf
RFA (2008) The Gallagher review of the indirect effects of bioufuels production, RFA 2008 https://www.unido.org/fileadmin/user_media/UNIDO_Header_Site/Subsites/Green_Industry_Asia_Conference__Maanila_/GC13/Gallagher_Report.pdf
RTRS (2010) RTRS standard for responsible soy production, http://www.responsiblesoy.org/quienes-somos/que-es-la-rtrs/
Scarlat (2011) Nicllae Scarlet, Jean-Francois Dallemand, recent developments of biofuels/bioenegy sustainability certification: a global overview. Energy Policy 39:1630–1646. 2011
SCOPE (2008) Biofuels: environmental consequences and interactions with changing land use http://www.eeb.cornell.edu/howarth/web/SCOPEBiofuels_home.html
Uchida (2007) Makoto Uchida, US and Brazilian diplomatic policies on ethanol which impacted on food policy, International Trade and Investment, Summer 2007, No.68, pp93–102 (in Japanese)
UN Energy (2007) Sustainable bioenergy: a framework for decision makers. http://www.fao.org/docrep/010/a1094e/a1094e00.htm
UNCTAD (2008) Making certification work for sustainable development: the case of biofuels, http://unctad.org/en/Docs/ditcted20081_en.pdf
UNEP (2009) Assessing biofuels, towards sustainable production and use of resources. International Panel for Sustainable Resource Management http://www.unep.org/pdf/assessing_biofuels.pdf
UNEP (2011) Towards a green economy http://www.unep.org/greeneconomy/greeneconomyreport/tabid/29846/default.aspx
USAID (2009) Biofuels in Asia an analysis of sustainability options https://www.cbd.int/doc/biofuel/USAID-biofuels-asia-2009-03.pdf
USEPA (2011) Biofuels and the environment: first triennial report to Congress, (Executive Review Draft) http://cfpub.epa.gov/si/si_public_record_report.cfm?dirEntryID=217443
World Bank a (2010) Biofuels markets, targets and impacts, July 2010, World Bank Policy Research Paper 5364 http://elibrary.worldbank.org/doi/pdf/10.1596/1813-9450-5364

Chapter 14
A Regional Perspective on Biofuels in Asia

Mark Elder and Shinano Hayashi

14.1 Introduction

In the beginning of the biofuel boom in the late 2000s, there were high expectations in many Asian countries that biofuels could enhance energy security, provide jobs, and reduce greenhouse gas (GHG) emissions. There were hopes that biofuels could be produced and consumed on a large scale and high expectations of significant biofuel trade. Some countries – particularly developing countries – hoped for biofuels to become a new major source of exports. For example, many in Indonesia hoped that their country could become the "Middle East of biofuels." Likewise, some developed countries, including Japan and some EU countries, hoped that significant biofuel imports, particularly from Southeast Asia, could diversity their energy sources. Thus, at that time, a regional perspective or strategy might have expected some countries (especially developing countries) to become major biofuel exporters and others (especially developed countries) to become major biofuel importers, with some potential interregional trade as well. Sustainability issues might be solved through a mechanism to apply sustainability standards.

This vision of a regional strategy or perspective assumed that significant land and other resources would be available to produce biofuel feedstocks on a reasonably large scale and in a sustainable manner. However, it generally has been very difficult to concretely identify large amounts of specific available land and assess whether adequate water is available, even before addressing sustainability issues. This chapter does not undertake a comprehensive study of available land and other resources, but rather reviews some existing efforts. It also considers the prospects for large-scale trade in biofuels to contribute to a major expansion of biofuel use.

M. Elder (✉) • S. Hayashi
Institute for Global Environmental Strategies (IGES),
2108-11 Kamiyamaguchi, Hayama, Kanagawa 240-0115, Japan
e-mail: elder@iges.or.jp

K. Takeuchi et al. (eds.), *Biofuels and Sustainability*, Science for Sustainable Societies, https://doi.org/10.1007/978-4-431-54895-9_14

Overall, this chapter concludes that large-scale increases in production are probably not realistic without large-scale diversion of land from other uses such as food production and without further pressure on the environment and other resources necessary for production, particularly water. To put it in more concrete terms, the current level of production of biofuels is generally modest, generally accounting for less than 5% of transport fuel in Asian biofuel-producing countries. Already at this level, the sustainability of biofuels has been questioned, although in some areas outside the region, such as the USA and Brazil, the share of transport fuel accounted for by biofuels is higher. While it could be conceivable to expand current levels to some extent, it seems clear that it is very difficult to expect biofuels to replace a large share of transport fuel such as 50% or even 30% or 20%.

Encouragement of smaller-scale production tailored to local conditions as a way to promote rural development and poverty reduction or as a way to address waste management issues may be more realistic. However, without large economies of scale of production, it will be difficult to reduce costs. In addition, there are various other challenges to the promotion of small-scale biofuels such as capacity of farmers; availability and cost of land, water, labor, and other inputs; and availability of markets for final outputs. If the main goal is to increase rural employment rather than energy security or GHG emissions reduction, then there may be other ways to accomplish this rather than through biofuels.

Sustainability standards and certification systems are one possible way to encourage the development of biofuels in a positive direction. However, while they may enable sustainable incremental production, they cannot create new land for biofuels; moreover, if they are to be effective, they should restrain the availability of new land by preventing excessive land use change from forests or food crops. The main efforts to implement sustainability standards have been taking place in Europe, where they will be required as part of the Renewable Energy Directive (RED), and they will apply to biofuel imports (Spiegel 2011). At the time of writing, there was no comparable initiative in East Asia, although a global voluntary initiative, the Roundtable on Sustainable Biofuels (RSB), has been developed with the participation of stakeholders from Asia and elsewhere, which could serve as the basis for an initiative in the region, and producers in the region could adopt the standards voluntarily. The RSB standard is still in the early stages of implementation, so it remains to be seen how effective it will be.

The rest of this chapter is organized as follows. First, the global feedstock requirements for biofuel production are examined. Second, the potential to expand biofuel production in East Asia is considered. Third, the limitations of second-generation biofuels are discussed. The fourth section explores the potential for trade to expand biofuel use in the Asian region. The fifth section considers the potential of sustainability standards, and the sixth section concludes.

14.2 Global Feedstock Requirements for Biofuel Production

This section discusses the global feedstock requirements for current biofuel production and uses them to estimate the potential for significantly expanded production. Already, globally, a significant share of the world's grain and vegetable oil production is being used to produce biofuel. However, this has only succeeded in replacing a small amount of liquid transport fuel.

Table 14.1 shows that overall, 11% of coarse grains and vegetable oils and 21% of sugarcane were used to produce bioethanol and biodiesel on average in 2008–2010. In 2020, this is expected to increase to 12–16% for grains and oils and 33% for sugarcane. Biofuels produced by these feedstocks accounted for 2.0% and 5.3% of diesel and gasoline use, respectively, on average in 2008–2010, and this amount is expected to increase to 3.8% and 8.8%, respectively, by 2020 according to Table 14.2.

Table 14.1 Present and future share of global coarse grains, vegetable oil, and sugarcane production used to produce biofuel

	2008–2010 average	2020 projection
Share of global production of coarse grains used to produce ethanol	11.0%	12.0%
Share of global production of vegetable oils used to produce biodiesel	11.0%	16.0%
Share of global production of sugarcane used to produce ethanol	21%	33%

Source: Calculated based on OECD-FAO Agricultural Outlook 2011–2020 (2011)

Table 14.2 Estimated share of global coarse grains, vegetable oil, and sugarcane production needed to significantly expand biofuel production

	2008–2010 average	2020 projection
Share of ethanol in global gasoline use (%) (energy shares)	5.3%	8.8%
Share of biodiesel in global diesel use (%) (energy shares)	2.0%	3.8%
Estimated share of global production of coarse grains needed for ethanol to replace 20% of global gasoline use	41.5%	27.3%
Estimated share of global production of vegetable oils needed for biodiesel to replace 20% of global diesel use	110.0%	84.2%
Estimated share of global production of coarse grains needed for ethanol to replace 50% of global gasoline use	103.8%	68.2%
Estimated share of global production of vegetable oils needed for biodiesel to replace 50% of global diesel use	275.0%	210.5%
Share of sugarcane in gasoline use	21%	33%
Estimated share of global production of sugar cane needed for ethanol to replace 20% of gasoline use	79%	75%
Estimated share of global production of sugarcane needed for ethanol to replace 50% of gasoline use	525%	434%

Source: Calculated based on OECD-FAO Agricultural Outlook 2011–2020 (2011)

Based on this, the amount of feedstock that would be needed to expand biofuels to account for 20% or 50% of gasoline and diesel use can be calculated. The result is that in order to offset 20% of gasoline and diesel use in 2008–2010, 41.5% of global coarse grain, 110% of vegetable oil, and 79% of sugarcane production would have been needed. To offset 50% of gasoline and diesel use, 103.8% of coarse grain, 275% of vegetable oil, and 500% of sugarcane production would have been needed.

Thus, the level of biofuel feedstock production and technology in 2008-2010 was not sufficient to replace 50% of either diesel or gasoline, since more than the entire amount of global feedstock production would have been required. Even to replace 20% of diesel or gasoline would have required much larger amounts of feedstock.

By 2020, it is projected that both feedstock production and the share of biofuel in gasoline and diesel will increase, because of increased productivity and stronger biofuel mandates. It appears that the percentage of biofuel feedstocks needed to replace 20% or 50% of gasoline and diesel would decrease to some extent. Nevertheless, a very large amount of feedstock would still be necessary to replace 20% of gasoline and diesel. To replace 50% of gasoline would still require more than half of the total production of coarse grains, and to replace 50% of diesel, even double the 2020 global vegetable oil production would not be enough. To be sure, technological advances and increases in yields may improve this situation to some extent, but the bigger picture is that there is a fundamental limit to how much biofuels can replace gasoline and diesel, considering that expansion of global biofuel production is constrained by a scarcity of farmland which will be needed to feed an increasing global population.

According to the FAO, to meet the needs of an expanding global population and adapt to changing consumption patterns, the world's food production will need to increase considerably over the coming decades, growing 70% above the level of 2009 by 2050 to feed an estimated additional two billion people. Much of this will need to be met by rising yields, although one study found that many biofuel feedstock crop yields have been overestimated, so that there might not be much room to increase them, and also rising yields may lead to environmental pressures (Johnston et al. 2009). The FAO says there is some room to expand biofuel feedstock production, but many of these potential new production areas are far from areas where biofuels would be consumed and not necessarily suited for the crops in the highest demand. Thus, most production growth would probably have to occur on existing agricultural land (FAO 2011).

Water shortages will also be a concern, and this issue was examined by FAO (2011). FAO's report carefully avoided concluding that there is not enough water for biofuels, but rather explained that there will be increased competition for water, as well as land, among different uses including food and fuel. A study by SEI examining the water energy and food nexus calculated that completely replacing fossil transport fuels would require 30 million barrels of ethanol and 23 million barrels of biodiesel per day, and only 10% of the required ethanol would require an additional 600 km^3 of water per year, which is much more than the global consumptive combined municipal and industrial water use (Hoff 2011, 19). Water is needed not only for the additional feedstock production but also for the fuel refining process. Thus, biofuels will compete for water as well as land. Water availability is a key global challenge (UNEP 2007) which is particularly severe for Asia (Kataoka and Shrestha 2010; de Fraiture et al. 2008). So future biofuel plans and targets will need to clearly indicate where the water will come from.

Therefore, it might be manageable to replace up to 10% of gasoline and diesel, although even this may be limited by sustainability constraints. However, there appear to be strong physical limitations to going much beyond this.

Of course, this calculation assumes the extrapolation of current technologies and production methods and does not take into account the possibility of yield or other productivity increases, so it may overestimate the amount of feedstock needed. Nevertheless, it is clear that productivity increases would have to be very substantial in order to make much of a difference. It may not be very realistic to expect this. The FAO cautions that while there is some room to increase crop yields, this involves certain environmental risks which would involve some difficulties in managing (FAO 2011).

14.3 Potential to Expand Biofuel Production in East Asia

In recent years, a number of Asian countries have established mandates or targets for biofuel use, as well as corresponding promotion policies. These targets have ranged from modest to ambitious as can be seen from Tables 14.3 and 14.4. China and India (bioethanol) and Indonesia and Malaysia (biodiesel) have been producing a significant amount of biofuels. Other countries in East Asia such as the Philippines (biodiesel), Thailand, and Vietnam (ethanol) also have been increasing their biofuel production, although the scale still remains modest. Most countries have found it very difficult to meet their targets due to difficulties in expanding production.

To what extent is it realistic to expect that countries might be able to meet their targets or otherwise significantly expand biofuel production, even aside from sustainability criteria? Assessing the physical potential for expansion of biofuel production is a very difficult exercise, since there is insufficient data in many cases. In particular, the reliability of land availability data is often questionable.

In most Asian countries, most land is already being used, or is a forest, or may not be very productive. Strong demand for biofuels will likely cause a shift in land use from food energy crops and may eventually cause deforestation. Although feedstock production technologies are advancing, there is limit to how much yields can increase. The use of agricultural waste has been suggested as a source of additional biofuel production, but this is sometimes already being used for other purposes or is providing ecosystem services. Collection of these wastes may be difficult or uneconomic. The use of marginal lands to grow nonfood crops that need little water such as *Jatropha* and others has also been suggested. However, while it is possible to grow crops such as *Jatropha* on wastelands with little water, yields will be low, and costs will be high if they are not irrigated or provided with fertilizer or planted on more fertile land. Moreover, marginal lands, sometimes called "wastelands," actually often are used by poor people with insecure land use rights.

Table 14.3 Future demand and production of bioethanol in the case study countries

Country	Target year	Scenario	Blending mandate/target	Gasoline demand in target year	Ethanol required	Production in target year				Total ethanol production	Target achieved?
						Current production	Ethanol from waste	From new land	From residue		
China	2020	Medium	10MMT	113,072	12,700	6686	5309	5905	10,778	28,678	Yes
India	2017	Low	20%	15,312	3062	2562	2690	215	4350	9817	Yes
Indonesia	2015	Low	10%	22,269	2227	212	856	563	660	2291	Y/N
Philippines	2011	Low	10%	4660	466	105	279	0	0	384	No
Thailand	2011	Low	10%	4893	489	408	434	330	0	1172	Yes
Vietnam	2020	Medium	500ML	6592	500	164	482	355	729	1730	Yes

Unit: million liters. *Source:* modified from USAID (2009)

Table 14.4 Future demand and production of biodiesel in the case study countries

Country	Target year	Scenario	Blending mandate/target	Diesel demand in target year	Biodiesel required	Production in target year		Total biodiesel production	Target achieved?
						Current production	From new land		
China	2020	Medium	2MMT	225,134	2400	355	224	579	No
India	2017	Low	20%	64,706	12,941	317	33	350	No
Indonesia	2010	Low	10%	11,593	1159	753	144	897	No
Malaysia	2010	Low	5%	6466	323	443	–	443	Yes
Philippines	2009	Low	2%	5967	119	211	–	211	Yes
Thailand	2012	Low	10%	17,338	1,734	48	236	284	No
Vietnam	2020	Medium	50ML	11,541	50	–	34	34	No

Unit: Million liters. *Source:* modified from USAID (2009)

The potential of several Asian countries to expand biofuel production to reach these targets has been analyzed by USAID (2009) which analyzed different possible scenarios for expanding their biofuel production, taking into account requirements for land and water. The study considered the options of shifting some land use from forest and food crop uses, increasing crop yields through the use of new technologies, use of crop residues, and use of marginal lands. The study considered both first-generation feedstocks, whose production increase may need to focus on underutilized land, and second-generation biofuel technologies using waste and residues from existing agricultural lands. Two scenarios considering crop mix, land use, and increasing yields were provided. The first scenario assumed expanding production of six crops using underutilized land, taking into account each country's specific resources. The second scenario assumed the use of five crops (maize, rice, sorghum, sugarcane, and wheat) to produce feedstock for second-generation bioethanol. Then, the scenarios were considered with each country's strategy for future biofuel production for approximately the next 10 years. The results of this study are summarized in Tables 14.3 and 14.4 above.

The USAID report optimistically concluded that for ethanol, around the year 2020, four of the six case study countries can probably meet their targets; one is borderline, and the other will fall somewhat short. However, for biodiesel, meeting the targets may be much more difficult, and only two of the seven countries are expected to do so. Moreover, the main reason why Malaysia and the Philippines are expected to meet their biodiesel production targets is because the targets are low.

The conditions necessary to reach the optimistic conclusion in the case of ethanol can be clearly seen from USAID's analysis. In all six case study countries, current production is less than the target, in some cases considerably less. Reaching the targets could be accomplished by significant expanded use of waste or residues and would not necessarily require new land.

Although the USAID study was able to identify some potential new land that could be converted to produce ethanol in each case study country except for the Philippines, the amount of land was not large. Presumably, this was the amount of land that USAID believed could be diverted without significant impact on food production or forests. Any significant increase in the amount of land used for biofuel feedstock production would cause concerns about land being converted from food production or forests.

In the case of biodiesel, the current production levels of the countries which are unable to meet their targets are generally significantly below the targets. To be sure, the USAID study did not consider the potential for waste to biodiesel or crop residues, so there could be some future potential for this. Nevertheless, the USAID study also did not identify significant new land that might be available for biodiesel production, at least not enough to enable these countries to meet their targets. The case of Indonesia is particularly important, since it was initially hoped that it could be a major exporter of biodiesel. In Indonesia, the main existing crop suitable for biodiesel is palm oil, for which Indonesia is one of the world's largest producers. However, palm oil is the main source of cooking oil in Indonesia, so its price and availability are very politically sensitive. Palm oil is also a major ingredient in many other products – including packaged foods – and India is a significant importer of palm oil for cooking. Therefore, any significant expansion of the use of palm oil for biodiesel would negatively impact its use as cooking oil, or in other products, or could encourage deforestation to make way for new palm oil plantations.

Table 14.5 presents a hypothetical scenario showing how much gasoline and diesel could be replaced if the entire production of selected feedstock crops in selected countries were to be entirely used for biofuels. For example, in the case of Indonesia, in 2008–2010, the country produced 272 million liters of biodiesel which accounted for 1.3% of diesel use (by energy share). Indonesia also produced 64 million tons of palm oil in 2005. USAID calculated that about 230 l of biodiesel can be produced from one ton of palm oil feedstock. Therefore, if Indonesia's entire production of 64 million tons of palm oil hypothetically could be converted to 14.7 billion liters of biodiesel, extrapolating from the share of diesel accounted by current biodiesel production, and assuming that all of the current biodiesel production is based on palm oil, the result is that converting all of Indonesia's palm oil to biodiesel would replace only about 70% of diesel fuel. This is admittedly a very rough, back of the envelope calculation. A number of factors could increase the potential replacement ratio, for example, if more crops were included or land productivity was higher. But the calculation is also conservative, in that it double counts the existing feedstock use, thereby overestimating the potential replacement ratio (possibly to a significant extent).[1] Overall, it gives an indication of the potential scale of biofuels in comparison to the use of liquid fossil fuels. It suggests that it may be quite difficult to expand crop-based biofuels to much more than 10% of liquid fossil fuels.

Tharakan et al. (2012) include an estimate of the potential for biofuels in the countries in the Greater Mekong Subregion (GMS). They note that available statistics optimistically suggest the potential to produce large amounts of biofuels. However, they conclude that the actual potential is much more modest if social and environmental risks are taken into account; moreover, land availability statistics typically are not very accurate in these countries. They note that while land availability currently is not a serious concern, expected increases in population and corresponding demand for food could generate increasing competition for land, and increasing risks of climate change and extreme weather are likely to adversely affect agricultural productivity.

Even in the more optimistic case, the ability of the GMS countries to generate significant exports is limited (see Table 14.6). In 2009, under the assumption that 10% of arable land could be used for biofuels, only Myanmar and Laos could have

[1] For example, in the case of Indonesia, all of the palm oil is assumed to be used to achieve the 70% replacement of biodiesel. However, some of the palm oil was already used to achieve the existing 1.3% replacement ratio, which was the basis for the extrapolation. This double counting is thus not very significant in the case of Indonesian biodiesel, but it might make more difference in the case of ethanol in Brazil. This is because a significant part of the sugar crop is already included in the current replacement ratio, which is already high at 47%. This calculation implies that about two thirds of the sugar crop is already used for ethanol, and converting the other one third would only push the replacement ratio up to 70%.

Table 14.5 Potential for biodiesel and ethanol to replace diesel and gasoline in selected countries

	Biofuel production	Domestic use	Fuel use	Share in fuel use (energy Sh.)	Share in fuel use (volume)	Net trade	Feedstock and date of data[a]	Total prod. feedstock[a]	Biofuel from feedstock[b]	Biofuel produced if 100% feedstock is used	Share of fuel use if 100% feedstock used
	Av 2008–2010 MN L	Av 2008–2010 MN L	Av 2008–2010 MN L	Av 2008–2010	Av 2008–2010	Av 2008–2010 MN L		MT	L of biofuel/ ton of feedstock	MN L	Share of domestic fuel use
				%	%						%
Biodiesel				(Share of diesel)							
Indonesia	369	272		1.3	1.7	98	Oil palm 2005	64	230	14,720	70.3
Malaysia	765	206		1.6	2.0	559	Oil palm 2005	76	230	17,480	135.8
USA	1658	909		0.3	0.4	748	Soybeans 2005	83	183	15,189	5.0
Brazil	1550	1550		2.7	3.4	0	Soybeans 2005	50	183	9150	15.9
Ethanol				(Share of gasoline)							
China	7189	7041	2024	1.8	2.6	148	Maize 2005	133	400	53,200	47.3
India	1892	2109	183	0.9	1.4	−217	Sugarcane 2005	232	70	16,240	79.9
US	42,857	44,663	42,338	5.3	7.7	−1806	Maize 2005	280	400	112,000	14.0
Brazil	26,091	22,589	21,061	47	57	3502	Sugarcane 2005	420	70	29,400	66.0

Calculated based upon OECD-FAO Agricultural Outlook 2011–2020 (2011)
[a]Rosamond et al. (2007)
[b]USAID (2009)

Table 14.6 Hypothetical share of domestic transport fuel demand that could be met through biofuels in GMS countries

	Bioethanol[a]		Biodiesel[b]	
	2009	2020	2009	2020
Cambodia	44	23	8	4
China (Yunnan and Guangxi)	73	40	4	3
Lao PDR	104	40	27	10
Myanmar	269	96	34	28
Thailand	13	3	1	0
Viet Nam	20	10	0	0

Source: Tharakan et al. (2012, 8)
Unit: percent
[a]Assumes bioethanol is produced from converting 10% of available land from wasted grain/crops
[b]Assumes biodiesel produced from converting 10% of available land

met all of their gasoline demand with bioethanol and have some leftover for exports. For Myanmar, this might be significant, as it would have been able to export ethanol equivalent to 169% of its gasoline consumption. However, the total quantity would not have been very significant in terms of total gasoline consumption of potential developed country importers. Moreover, by 2020, due to expected rising fuel consumption in the GMS countries, only 3–40% of gasoline demand might be able to be met by domestic production and 96% in the case of Myanmar. For biodiesel, the potential to replace diesel fuel is much less (Tharakan et al. 2012). Thus, while the authors find that biofuels could make some contribution to domestic transport fuel in GMS countries, it seems clear that there will not be sufficient capacity for any of these countries to become leading suppliers to developed countries. Like Indonesia, the GMS countries have no prospect of becoming the Middle East of biofuels.

Even in the case of production for domestic use, Tharakan et al. (2012) observe that the extent to which this potential can be realized depends on various factors, including the type of production system that is used. GMS countries are subject to similar environmental and social constraints as in other areas, in particular potential food-fuel conflicts. They note that large-scale industrial plantations – which would be necessary for large-scale exports – would be particularly problematic. They conclude that smaller-scale production based on surplus land, nonfood crops, and smallholder-based production is more realistic (Tharakan et al. 2012, 413).

14.4 Limitations of Second-Generation Biofuels

It has been hoped that so-called second-generation or advanced biofuels can overcome the limitations of first-generation biofuels based on conventional agricultural feedstocks. However, in practice, these are also subject to a variety of physical limitations limiting the scope of their potential, besides waiting for technical advances. For example, the IEA has estimated that 10% or 25% of the global forestry and agricultural residues in 2007 could have produced enough biodiesel and ethanol to provide 4.2–6.0 to 10.5% of current transport fuel demand, respectively (IEA 2010, 9). To be sure, this could be an important partial contribution, but it is still not enough to serve as the main source for transport fuel.

Many second-generation biofuels – including switchgrass, algae, etc. – require land and water, just as first-generation ones do. In many cases, waste from forests and agriculture perform ecosystem services such as returning nutrients to the soil, so there is a fundamental limitation on how much these resources can be exploited. In particular, this "waste" is often used by small-scale farmers in the region for fertilizer, so if it is used to produce biofuel, then farmers may be forced to use more conventional fertilizers (Elder et al. 2008, 13).

14.5 Biofuel Trade: A Scramble for Biofuels?

Much of the existing discussion on biofuel trade has focused on criticizing the common practice of using protectionist policies to provide advantages to domestic biofuel producers and estimating the resulting economic inefficiencies and costs. Much less attention has been paid to the underlying logic of biofuel trade and its connection to sustainability issues.

At the beginning of the biofuel movement, most interested countries aimed to nurture domestic producers. However, while some intended to be mainly self-sufficient, others, particularly in Europe and Japan, realized the impossibility of self-sufficiency, and intended to supplement domestic production with imports, partly to enhance energy security, but also partly as a way to reduce greenhouse gas emissions. Soon, it became apparent that some countries which intended to be self-sufficient could not, and overambitious targets would need to be met through imports. Still others had ambitions to be major biofuel exporters. For example, some in Indonesia hoped for their country to become the "Middle East of biofuels," and indeed, at one time, the EU had hoped to import significant quantities of biofuels from Indonesia or other Southeast Asian countries.

Now, it has become apparent that many or most countries will not be able to achieve their targets, even relatively modest ones, and few countries will be able to develop large-scale exportable surpluses beyond their domestic requirements. There is not likely to be any "Middle East of biofuels," in East Asia, not even Indonesia.

In fact, only a small share of biofuel production has been traded globally, about one-tenth, as can be seen from Figs. 14.1 and 14.2. Moreover, an OECD-FAO study suggested that increasing global biofuel production will not necessarily lead to increased global biofuel trade in the future either.

Of course, biofuel trade exists and will continue, but it will ebb and flow based on marginal supply and demand differences and price fluctuations among feedstocks and between biofuels and fossil fuels, etc. Differences in biofuel and feedstock trade protection policies, biofuel and fossil fuel promotion policies and subsidies, and blending mandates will also induce trade. In particular, countries

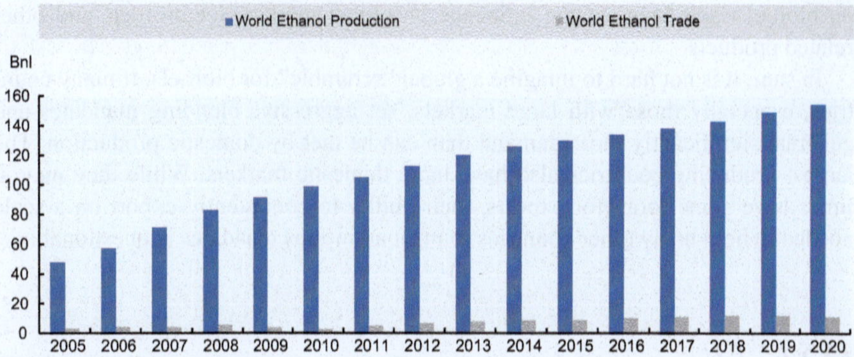

Source: OECD and FAO Secretariats

Fig. 14.1 Development of the world ethanol market

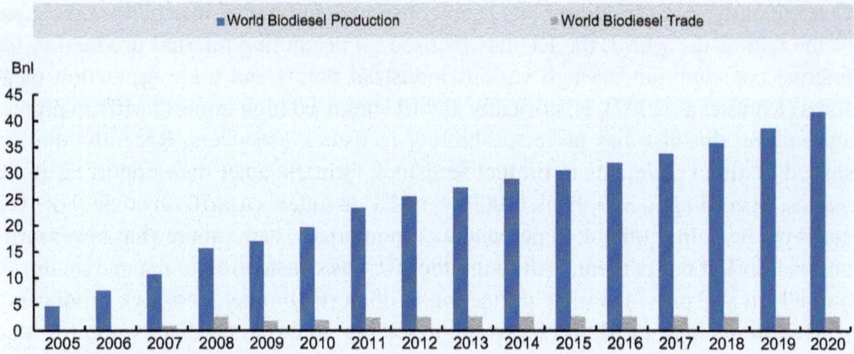

Source: OECD and FAO Secretariats

Fig. 14.2 Development of the world biodiesel market

with strong biofuel mandates but inadequate domestic production are likely to attract imports unless domestic producers are granted trade protection. Conversely, in this situation, if trade protection for domestic producers is strong enough, then the biofuel mandate would remain unmet if the domestic producers can or will not increase production.

Some countries, such as the USA or Brazil, may at times export some biofuels which cannot be absorbed into their domestic markets, but the available quantities are not likely to be enough to enable major increases in biofuel utilization mandates in many countries simultaneously. Moreover, the US could alternate from being a net exporter to a net importer, so export volumes could be unstable. The EU and USA are particularly significant, since they account for a majority of the world's biodiesel production, and the USA and Brazil account for 89% of global bioethanol production, although the amount exported is much smaller. Their policy decisions on biofuel trade have a large influence on global markets for biofuels and other related products.

In sum, it is not hard to imagine a global "scramble" for biofuels, if many countries, especially those with large markets, set aggressive blending mandates and generate significantly more demand than can be met by domestic production. The largest producing countries also have large domestic markets. While they may at times have some room for exports, their ability to consistently export on a scale needed to help many other countries to meet ambitious mandates is questionable.

14.5.1 EU

The EU's share of global biodiesel production in 2009–2010 was approximately 65% (European Biodiesel Board). The production of biodiesel has been heavily subsidized because its production cost has been higher than that of fossil fuels. Consequently, EU's biodiesel has been consumed and traded internally. Like most of the rest of the world, the EU has focused on promoting internal production for internal consumption through various industrial policy and trade protection measures (Kutas et al. 2007). Historically, the EU has used high import tariffs to protect agriculture; this also has protected biofuel feedstock producers. Recently, the EU shifted to direct payments to biofuel feedstock farmers rather than import tariffs or quotas. According to Swinbank (2009), "the EU maintains a tariff on ethanol of 10.2 euros per hectoliter (about 45 percent at current prices) and a somewhat lower tariff on biodiesel of 6.5 percent." Although the EU was considering to expand its use of biofuels, it still provided trade protection to domestic biofuel feedstock producers.

Despite its high tariff and focus on protecting domestic producers, the EU also set a high biodiesel blending mandate. This implied that imports would also be required since EU production would not be able to produce enough to meet the mandate (and/or that some EU land would have to be diverted from food to fuel crops). This strict blending mandate reflected not only a desire for cleaner energy but was also intended to enhance energy security and supply diversification.

The European Commission's strategy, "An EU Strategy for Biofuels," was published in 2006 (EU 2006). It stated that stimulating trade opportunities and supporting biofuel producers in developing countries were key elements of the strategy. The strategy aimed to secure biofuel supplies from developing countries and to facilitate the production of crude vegetable oil for bioenergy. The EU intended that the biofuels from the high blending mandate would be complemented with imports from countries like Malaysia and Indonesia. The EU's main environmental concern at that time was unilateral reduction of greenhouse gasses, and it had not considered the potential for sustainability issues arising from production in the expected exporting countries (Jank et al. 2007). In fact, the Netherlands was the biggest market for refining and combustion of palm oil at the time (Greenpalm.org 2011). The production of "green electricity" had the potential to boost demand for palm oil by more than 1,000,000 MT annually; nevertheless, the Dutch government stopped its subsidy for using palm oil for electricity, because of negative publicity regarding the sustainability of palm oil production in Malaysia and Indonesia. EU leaders became sensitive to criticism that their biofuel promotion policy might be leading to deforestation and higher food prices (Harrison 2008). In the response to this, the EU considered revising or reinterpreting the standard (Euractiv.com 2008; Reuters 2010). The EU relaxed the biofuel blending mandate and started working on a possible biofuel certification considering environmental impacts (Al-Riffai et al. 2010).

In 2009, the EU adopted the Renewable Energy Directive (2009/28/EC) which established a target of 10% of the energy used for transport within the EU to come from renewable sources. It was understood that a significant portion of this would come from biofuels. However, the Directive also imposed sustainability requirements (Spiegel 2011). In order to count toward the target, biofuels must not be produced at the expense of primary forests or carbon-rich soils such as peatlands and must demonstrate a savings of greenhouse gasses of at least 35% compared to fossil fuels (RSB 2011, 3). Therefore, while the risk of large-scale imports of unsustainably produced biofuels was reduced, the EU will probably remain a major importer. It is not likely to become a major exporter of biofuels and instead will compete for imports with other countries with high blending mandates.

As of 2012, the biofuel share of transport fuel reached only 4.65%, far short of the 10% target, and much of that had to be imported (USDA 2011). The EU probably will not have sufficient feedstock production capability by 2020 to reach its expected 6.6%, biodiesel target blend, so net biodiesel imports are expected to be more than 2 billion liters, and about 2.3 billion liters of net ethanol imports will be needed to reach the expected level of 8.2%, ethanol blend. Between 2008 and 2010, the EU imported on average 1.5 billion liters of ethanol and 1.6 billion liters of biodiesel (OECD/FAO 2011). In the meantime, vegetable oils for human consumption and industrial use (e.g., cosmetics) have to be imported, so imports of biofuels or vegetable oils for use as biodiesel will compete with these uses (Jank et al. 2007).

14.5.2 USA

In the USA, approximately 95% of bioethanol is made from maize, and 90% of biodiesel is made from soybeans. Although ethanol from maize in the USA is generally more costly to produce than sugarcane used for ethanol in Brazil, it was cheaper in 2000 when sugar prices in Brazil hit their peak, while US maize prices dropped. US production and consumption of ethanol have accelerated in recent years. In 2008, nearly 30% of maize produced in the USA was used for ethanol, and 20% of soybean production was used for biodiesel. High blending targets made the USA the world's largest ethanol importer, since it was not able to meet these targets solely through domestic production. In 2006, the USA accounted for more than half of global ethanol imports, and Brazil accounted for more than half of US imports.

The USA also has imported ethanol from Caribbean Basin Initiative (CBI) countries which can enter duty-free. Although the USA has adopted policy measures to increase biofuel production, it has not been able to keep pace with the rapid increase in biofuel consumption (Jank et al. 2007). Therefore, if the USA maintains its current blending mandate, it might be sufficient to increase imports of ethanol gradually from the small CBI countries. However, if the mandate increases further, it would be necessary to import larger quantities of ethanol from countries such as Brazil with larger production capacities.

Elobeid and Tokgoz (2006) conducted a study simulating the removal of US import tariffs on ethanol. This study estimated that removing the tariff would increase world's ethanol prices by 24% and sugar prices by 1.8%, and it would decrease maize prices by 1.5%. In the USA, ethanol prices would fall by 14%, since cheap imports from Brazil would displace imports from Caribbean countries, and US consumption would increase by 4%. In Brazil, ethanol consumption would drop by 3%, and ethanol exports would increase by 64%.

The USA has been a major importer of biofuels in the past, and imports could potentially increase if the US consumption continues to outpace production, and the blending mandate becomes more aggressive. However, this potential may be moderated by increasing the productivity of maize production and continued political pressure to maintain trade protection for domestic producers. Although the USA also exports some biofuels, it seems unlikely that it could become a major consistent supplier to many other countries, at least not without significantly diverting more of its food crops to use for biofuel production.

14.5.3 Brazil

At the time of writing, Brazil was the world's largest bioethanol consumer, as well as one of the most efficient and low-cost producers. Brazil had neither production subsidies nor import tariffs on ethanol. Compared to corn-based ethanol produced by the USA, Brazil's sugarcane-based ethanol production had a much higher productivity, and Brazil's cheap labor was suitable for labor-intensive sugarcane production. Brazil also had abundant water, which is essential for large-scale production.

With these advantages, Brazil has been the world's most competitive producer of bioethanol. Brazil produced 22,100 million liters of bioethanol in 2008. According to the OECD-FAO Agricultural Outlook (2008), projected bioethanol production in 2017 will increase 83.3% from 2008 to a total of 40,500 million liters. Overall, Brazil is the sole country with the ability to potentially export a large amount of bioethanol in 2017 according to projections by OECD-FAO and FAPRI (2010).

Nevertheless, there are limits to Brazil's ability to export biofuels to the world. On one hand, from 2008/2010 to 2020, Brazil's production of ethanol is expected to rise from 26 to 50 billion liters. On the other hand, much of the projected increase in production is expected to be consumed domestically, as the replacement of gasoline increases from 47% to 75%. So only about 9.6 billion liters would be exported in 2020, although Brazil still would be the world's largest exporter by far. This represents about one fifth of Brazil's total production and 6.2% of expected global production of 155 billion liters. Incidentally, the USA is expected to be by far the largest importer, and the expected amount of 9.5 billion liters is nearly equal to Brazil's entire export amount. The EU is expected to import another 2.4 billion liters (Table 14.7).

14.5.4 Possibility of Large-Scale Trade

The fundamental limitations of large-scale trade are illustrated by the OECD/FAO projections (OECD/FAO 2011, 92). Out of the total global production of 91.6 billion liters of ethanol in 2008–2010, only about 3.8 billion liters were exported, and of this amount, about 3.3 billion liters were imported by the USA and the EU, leaving little available for other areas. In 2020, global ethanol production is expected to grow significantly to about 155 billion liters, but only 11 billion liters are expected to be exported, mostly by Brazil (9.7 billion liters) and mostly to the USA (9.5 billion liters). In Asia, the largest ethanol exporters in 2020 are expected to be China (1,200 million liters), Thailand (509 million liters), and the Philippines (153 million liters), while the major importers are expected to be Japan (769 million liters), India (614 million liters), and Malaysia (11 million liters) (OECD/FAO 2011, 92) (See Table 14.7).

For biodiesel, the scale of expected trade is even smaller. Total global biodiesel production in 2008–2010 of 17.6 billion liters was dominated by the EU (9.2 billion liters), the USA (1.7 billion liters), Argentina (1.6 billion liters), and Brazil (1.6 billion liters). Total global exports accounted for only 2.5% of total production; the main exporters were Argentina (1329 million liters), the USA (748 million liters), and Malaysia (559 million liters), with the EU being the largest importer (1.6 billion liters) (OECD/FAO 2011, 93).

In sum, there is no clear potential source of large-scale exports that would be needed if all Asian countries were to simultaneously increase their blending mandates significantly beyond domestic production capabilities. The world cannot rely on Brazil alone. If a sufficient number of countries in East Asia or elsewhere were to significantly increase biofuel blending mandates, it would likely cause a global biofuel supply shortage. Globally there would be pressure to shift more land to fuel crops and increase the risk of non-sustainable production practices.

It is worth mentioning the various barriers to biofuel trade (see Table 14.8). One of the main determinants of biofuel trade is the relative prices of inputs such as feedstocks and competing fossil fuels. These relative prices fluctuate considerably and contribute to the volatile nature of biofuel trade. Trade protection tends to be high, as countries want to promote their domestic industries. Biofuels also suffer from high transportation and insurance costs. All of these factors tend to discourage biofuel trade.

Generally, trade is considered to increase efficiency. However, this may not necessarily always be the case for biofuels, if the trade is motivated by strong blending

Table 14.7 Biofuel projections (ethanol)

	Production (MN L)		Growth (%)[a]	Domestic use (MN L)		Growth (%)[a]	Fuel use (MN L)		Growth (%)[a]	Share in gasoline-type fuel use (%)				Net trade (MN L)[b]	
										Energy shares		Volume shares			
	Average 2008–2010 est.	2020	2011–2020	Average 2008–2010 est.	2020	2011–2020	Average 2008–2010 est.	2020	2011–2020	Average 2008–2010 est.	2020	Average 2008–2010 est.	2020	Average 2008–2010 est.	2020
North America															
Canada	1493	2359	3.08	1530	2408	0.57	1324	2202	0.66	2.2	3.4	3.3	5.0	−48	−49
USA of	42,857	63,961	1.89	44,663	73,474	3.32	42,338	70,484	4.13	5.3	8.4	7.7	12.1	−1806	−9514
which second generation	3	4368	–	–	–	–	–	–	–	–	–	–	–	–	–
Western Europe															
EU(27) of	5651	16,316	10.50	7186	18,690	7.31	4687	16,173	8.09	2.3	8.2	3.4	11.8	−1536	−2374
which second generation	0	1626	–	–	–	–	–	–	–	–	–	–	–	–	
Oceania developed															
Australia	299	492	0.75	299	492	0.75	299	492	0.75	1.0	1.6	1.5	2.3	0	0
Other developed															
Japan of	307	946	13.28	704	1715	5.81	90	1687	18.26	0.0	0.0	0.0	0.0	−398	−769
which second generation	0	593	–	–	–	–	–	–	–	–	–	–	–	–	–
South Africa	384	421	0.44	93	47	0.07	0	0	4.62	0.0	0.0	0.0	0.0	291	374
Sub-Saharan Africa															
Mozambique	25	59	6.17	21	29	0.56	0	9	1.48	0.0	3.3	0.0	4.8	4	29
Tanzania	29	55	7.14	33	52	5.97	1	19	37.15	0.1	2.7	0.2	4.0	−4	3

Latin America and Caribbean															
Argentina	303	470	2.20	240	402	0.97	110	272	1.47	1.6	3.4	2.3	5.0	63	68
Brazil	26,091	50,393	5.93	22,589	40,695	5.15	21,061	38,383	7.28	47.3	67.1	57.2	75.3	3502	9698
Columbia	310	587	5.63	353	385	−1.20	315	347	−1.33	4.5	5.6	6.6	8.1	−44	202
Mexico	64	90	2.29	168	275	2.29	0	0	–	0.0	0.0	0.0	0.0	−104	−184
Peru	71	217	2.55	25	175	1.47	20	174	1.48	1.1	8.2	1.7	11.7	46	41
Asia and Pacific															
China	7189	7930	0.71	7041	6685	0.18	2024	2975	4.34	1.8	1.5	2.6	2.3	148	1,246
India	1892	2204	1.78	2109	2818	1.48	183	800	1.48	0.9	3.0	1.4	4.5	−217	−614
Indonesia	210	248	0.99	169	168	0.15	0	0	6.77	0.0	0.0	0.0	0.0	41	80
Malaysia	66	74	0.80	87	85	0.09	0	0	5.38	0.0	0.0	0.0	0.0	−21	−11
Philippines	118	603	12.74	263	450	3.49	193	350	−0.30	2.1	3.0	3.1	4.4	−144	153
Thailand	672	2111	9.32	599	1602	8.72	424	1389	4.54	3.8	11.2	5.6	15.9	73	509
Turkey	64	88	0.98	108	142	3.43	50	87	5.23	0.6	0.9	1.3	−44	−54	
Viet Nam	150	423	4.75	95	334	14.84	8	255	25.87	0.1	3.5	0.2	5.1	55	90
Total	*91,657*	*154,962*	*3.98*	*91,821*	*155,983*	*3.95*	*73,742*	*136,123*	*4.45*	*5.3*	*8.8*	*7.7*	*12.6*	*3,792*	*11,012*

Source: OECD and FAO Secretariats

– Data not available

[a]Least-squares growth rate

[b]For total net trade exports are shown

Table 14.8 Bioethanol and biodiesel import tariffs

Country	Ethanol ($/L)[a]	Biodiesel ($/L)[b]
Australia	$0.24	na
Brazil	$0.70	na
Canada	$0.50	na
EU	$0.10	Ad valorem duty of 6.5%
Japan	$0	na
USA	$0.14	$ 0.26

[a]IEA
[b]Swinbank (2009)

mandates or subsidies, which may instead worsen market imbalances caused by highly distortionary biofuel promotion policies.

Expanded trade, including through liberalization of trade restrictions, does not necessarily enhance sustainability either. Particularly in most Asian countries, there appears to be little room to increase land devoted to biofuel production without deforestation or shifting from the production of food crops, so a significant increase in production may be associated with significant risks of reduced sustainability. Therefore, from the standpoint of biofuel-producing countries, to increase exports significantly would risk reducing sustainability of biofuel production by stimulating it beyond the limits of sustainability. Likewise, if the sustainability of production in an exporting country were to decline, the sustainability of the energy use of the importing country would also worsen rather than increase. This problem was illustrated in the debate over the EU's blending mandate in the 2000s. At that time, it was expected that imports from Indonesia and other places would help it to meet ambitious mandates, but as an increasing number of studies raised serious doubts about the policy's likely impacts on sustainability, particularly the risk of deforestation, the EU suspended and finally canceled the policy's implementation and shifted to a more modest mandate combined with sustainability standards. By the same token, trade protection for domestic biofuel producers might enhance sustainability by restraining the scale of production to more sustainable levels.

14.6 Sustainability Standards

Various efforts have been made to develop sustainability standards and certification systems for biofuels in order to enhance their sustainability (Dam et al. 2008). They give producers a chance to demonstrate it on a case-by-case basis. However, by themselves, they cannot create new land or other resources such as water. Moreover, it is too early to assess their potential effectiveness, as they have not been extensively implemented.

At the time of writing, there were no officially recognized global or regional biofuel sustainability standards. An international dialogue on sustainability criteria and the development of transparent and harmonized standards and certification schemes was held through various frameworks. The Roundtable on Sustainable Palm Oil (RSPO) and Roundtable on Responsible Soy (RTRS) are commodity-based initiatives with criteria for certification. The Better Sugarcane Initiative is another roundtable initiative focusing on biofuel feedstocks (Elder et al. 2008). The Roundtable on Sustainable Biofuels has developed a global comprehensive, voluntary, multi-stakeholder initiative including standards as well as a certification system. At the same time, the G8 countries created the Global Bioenergy Partnership (GBEP) which also consisted of public, private, and civil society stakeholders in a joint commitment to promote bioenergy for sustainable development. GBEP is potentially the most important one, since it includes governments as members, but it was "not willing to develop an additional standard and certification scheme" (Scarlat and Dallemand 2011); therefore, GBEP was expected to reach consensus on biofuel sustainability as a meta-standard. The UK has established its own Renewable Transport Fuel Obligation (RTFO).

The "roundtable approach" provides opportunities to develop certification systems supported by a wide range of stakeholders. Nonetheless, as the criteria developed by those roundtables are only voluntary commitments, this approach will be effective only if all stakeholders actually follow the criteria. Another concern is the motivation of the participants. Some NGOs argued that the roundtables provide some governments an excuse not to take stronger, more direct measures to protect the environment and vulnerable populations (Reuters 2007).

There are two main motivations for sustainability standards. The first is sustainability concerns, which might lead countries to develop their own standards, partly aimed at avoiding imports of unsustainably produced biofuels. The second is for export promotion, to make domestically produced biofuels more attractive to customers who are concerned about sustainability. For developing countries, the main concern may be standards in advanced countries that imports are required to meet. Establishing a reliable and robust but low-cost certification system is a major challenge. Other constraints for certificate and standard systems include implementation costs, physical and human capacity, and monitoring costs. Standards and certification systems may be particularly challenging for small producers, who may need assistance to be able to comply with these schemes.

One key issue for implementing biofuel sustainability standards/certificates is how to attract the participation of producers. Some producers may adopt the standard hoping to charge a higher price to environmentally conscious consumers, but many will adopt it only if it is required by government regulations or by customers.

However, there is no recognized global or regional standard. The EU has adopted a requirement for certification, but it has also encouraged competition among certification systems. The Roundtable on Sustainable Biofuels (RSB) has developed a global standard through multi-stakeholder dialogue, but it is voluntary. More time is needed to see how it will develop.

There are currently no efforts to develop regional sustainability standards in East Asia. The RSB standard could serve as a basis for one, as it was developed with input from Asian stakeholders, and producers have the option of adopting it voluntarily.

14.7 Conclusion

Overall, current modest levels of biofuel use in Asian countries may be feasible, but it is likely to be practically quite difficult for biofuels to account for a large share of transport fuel use of 20% or more. Some Asian countries established blending mandates of 5–10% of transport fuel, but even these levels will be very difficult to meet with domestic production.

It is not clear where new large-scale production can come from. Waste or agricultural residue may be potential sources, but there are various problems in accessing them. Some new land may be available, but it is not always clear where, and even if land can be found, it is not likely to be on a very large scale. In order to devote large-scale additional land to biofuel production, it would probably be necessary to divert food production or convert forestland. Productivity gains may help some, but many Asian countries may still experience population increases in the coming years, and some land and agricultural productivity increase will need to be devoted to increasing food crops and living space.

Some countries may need to meet their targets through imports, if the targets are strictly enforced. But it is not necessarily clear where large-scale imports could come from. The major global producers, the USA and Brazil, also reserve the bulk of their production for domestic use, and export volumes are unstable. Brazil's export potential is not necessarily enough to meet every country's blending mandate shortfall. Most countries are not strictly enforcing their blending mandates, so a "scramble" for imports has not occurred. But if many countries were to simultaneously enforce strict mandates, a scramble for imports could result.

Sustainability standards could be useful to discourage unsustainable production practices. The standards would be more effective if they are mandated by governments and need a robust certification system. However, standards themselves cannot create additional land or other resources.

Smaller-scale production based on individual local circumstances to promote rural development or address waste problems may be more realistic. However, without scale economies, the costs will be relatively high, and the contribution to energy security and GHG reduction will be modest. In any case, it seems better for each country to pursue its own strategy tailored to its individual circumstances and local conditions.

References

Al-Riffai P et al (2010) European Union and United States biofuel mandates: impacts on world markets. Inter-American Development Bank, Washington, DC

Dam JV et al (2008) Overview of recent developments in sustainable biomass certification. Biomass Bioenergy 32:749–780, IEA Bioenergy Task 40 special issue

de Fraiture C, Giordano M, Liao Y (2008) Biofuels and implications for agricultural water use: blue impacts of green energy. Water Policy 10(Supplement 1):67–81

Elder M et al (2008) Prospects and challenges of biofuels in Asia: policy implications. In: Climate change policies in the Asia-Pacific: re-uniting climate change and sustainable development. Institute for Global Environmental Strategies (IGES), Hayama, pp 105–131

Elobeid A, Tokgoz S (2006) Removal of U.S. ethanol domestic and trade distortions: impact on U.S. and Brazilian ethanol markets. Working Paper 06-WP 427. Center for Agricultural and Rural Development Iowa State University

EU (2006) An EU strategy for biofuels

Euractiv.com (2008) Biofuels for transport, 11 April. http://www.euractiv.com/transport/biofuels-transport-linksdossier-188374 . Accessed 14 May 2012

FAO (2011) The state of the world's land and water resources for food and agriculture (SOLAW): managing systems at risk, summary report. Food and Agriculture Organization of the United Nations, Rome and Earthscan, London. http://www.fao.org/docrep/015/i1688e/i1688e00.pdf (Accessed May 14, 2012)

Food and Agricultural Policy Research Institute. FAPRI (2010) U.S. and world agricultural outlook. Iowa State University 2010. http://www.fapri.iastate.edu/outlook/2010/. Accessed 12 July 2011

Greenpalm.org (2011) RSPO supports Dutch industry's call on the EU to abolish import duty on sustainable palm oil. http://www.greenpalm.org/en/blog-press/blog/rspo-supports-dutch-industrys-call-on-the-eu-to-abolish-import-duty-on-sustainable-palm-oil. Accessed 12 May 2012

Harrison P (2008) EU backs away from biofuel goal. Reuters, 5 July. http://uk.reuters.com/article/2008/07/05/us-eu-energyidUKL0542894820080705? sp=true. Accessed 14 May 2012

Hecht AD et al (2009) Good policy follows good science: using criteria and indicators for assessing sustainable biofuel production. Ecotoxicology 18:1–4

Hoff H (2011) Understanding the nexus. Background paper for the Bonn 2011 conference: the water, energy, and food security nexus. Stockholm Environment Institute, Stockholm. http://www.water-energy-food.org/documents/understanding_the_nexus.pdf. Accessed 25 May 2012

International Energy Agency (IEA) (2010, February) Sustainable production of second-generation biofuels: potential and perspectives in major economies and developing countries. Information Paper

Jank MJ et al (2007) EU and U.S. policies on biofuels: potential impacts on developing countries. The German Marshall Fund, Washington, DC

Johnston M et al (2009) Resetting global expectations from agricultural biofuels. Environ Res Lett 4(014004):1–10

Kataoka Y, Shrestha S (2010) Promoting wise use of water: application of economic instruments. In: Institute for Global Environmental Strategies (IGES), White Paper III. Hayama, pp 149–170

Kojima M, Mitchell D, Ward W (2007) Considering trade policies for liquid biofuels. World Bank, Energy Sector Management Assistance Program (ESMAP), Washington, DC, p 141

Kutas G et al (2007) Biofuels—at what cost? Government support for ethanol and biodiesel in selected OECD countries. The Global Subsidies Initiative (GSI) of the International Institute for Sustainable Development (IISD), Geneva

OECD/FAO (2008) OECD-FAO agricultural outlook 2008–2017. OECD Publishing and FAO

OECD/FAO (2011) OECD-FAO agricultural outlook 2011–2020. OECD Publishing and FAO

Reuters (2007) Corrected: green group wary of plans for "eco-friendly" palm

Reuters (2010) EU report signals U-turn on biofuels target

Roundtable on Sustainable Biofuels (RSB) (2011) Biofuel for thought. 2(3) September. http://rsb.epfl.ch/files/content/sites/rsb2/files/Biofuels/Newsletters/RSB_Newsletter-September%202011.pdf. Accessed 5 Apr 2012

Scarlat N, Dallemand JF (2011) Recent developments of biofuels/bioenergy sustainability certification: a global overview. Energy Policy 39(3):1630–1646. https://doi.org/10.1016/j.enpol.2010.12.039

Spiegel (2011) Chaos at the pumps: German consumers are wary of new E10 biofuel

Swinbank A (2009) EU support for biofuels and bioenergy. 'Environmental sustainability' criteria, and trade policy. International Centre for Trade and Sustainable Development. Issue paper No. 17
Pradeep Tharakan, Naeeda Crishna, Jane Romero, and David Morgado. Biofuels in the Greater Mekong Subregion: Energy Sufficiency, Food Security, and Environmental Management. Asian Development Bank. ADB Southeast Asia Working Paper Series. No. 8. January.
UNEP (2007) Global environment outlook GEO4: environment for development
USAID (2009) Biofuels in Asia: an analysis of sustainability options
USDA Foreign Agricultural Service (2011) GAINS report. EU-27 annual biofuels report, no. NL1013, June 24

Chapter 15
National Strategy Options for Japan

Osamu Saito

15.1 Introduction

The introduction and diffusion of biofuel industry have been promoted in many developed countries including Japan, which has established concrete mandates with numerical targets for both bioethanol and biodiesel. Table 15.1 shows changes to the biofuel introduction targets in Japan. In response to government requests to achieve the GHG emission reduction goals of the Kyoto protocol, the Petroleum Association of Japan has agreed to blend 840,000 kl/year of bio-ETBE (ethyl tertiary-butyl ether), equivalent to 210,000 kl of crude oil, into gasoline starting in fiscal year (FY) 2010. This blended bio-ETBE gasoline has been sold as "biogasoline," and the number of service stations selling it has increased from 50 in 2007 to 3210 in 2012. On the other hand, Japan's Ministry of the Environment (MOE) has been promoting a strategy to accelerate the use of biomass energy by supplying E3 gasoline, a blend of gasoline with 3% bioethanol. Demonstration projects for E3 have been conducted in Osaka, Tokyo, and Okinawa, but the amount of E3 gasoline sold in 2010 remained approximately 28,000 kl.

A number of studies have evaluated how achieving these mandates can contribute to reductions in GHG emissions and how the expansion of biofuel production can affect food security. However, there are few studies focusing on the interlinkages between different impacts, including trade-offs and synergies among different types of impacts. This chapter quantitatively assesses various environmental impacts by expanding biofuel production and ethanol usage and analyzes the interlinkages among different impacts under several options for introducing biofuel in Japan. We use three indicators for this analysis, life-cycle carbon footprint ($LCCO_2$), water footprint (WF), and ecological footprint (EF), by considering feedstock types,

O. Saito (✉)
United Nations University Institute for the Advanced Study of Sustainability (UNU-IAS),
5-53-70 Jingumae, Shibuya 150-8925, Tokyo, Japan
e-mail: saito@unu.edu

K. Takeuchi et al. (eds.), *Biofuels and Sustainability*, Science for Sustainable Societies, https://doi.org/10.1007/978-4-431-54895-9_15

Table 15.1 Changes to biofuel introduction targets in Japan

April 2005	The plan for achieving the Kyoto protocol target (approved by the Cabinet on April, 28 2005) identified 3080,000 kl crude oil equivalent of biomass thermal energy use including 500,000 kl crude oil equivalent of liquid biofuel for transportation, which is equivalent to approximately 0.6% of the total liquid fuel for transportation (86,000,000 kl)
March 2006	New biomass Nippon strategy has also set the target of introducing 500,000 kl crude oil equivalent of liquid biofuel for transportation
May 2006	New national energy strategy has set the target to reduce petroleum dependency of transportation sector from 98% in 2000 to 80% by 2030
November 2006	Prime Minister Shinzo Abe directed the development of a road map to expand the domestic biofuel production up to 6000,000 kl, which is equivalent to 10 % of the annual domestic gasoline consumption
November 2010	A new law on nonfossil energy use and effective use of fossil energy resources by energy suppliers was enacted, and its public notice (No. 242)[a] indicated the following targets with respect to bioethanol usage
	Bioethanol usage targets from FY 2011 to FY 2017:
	FY 2011: 210,000 kl crude oil equivalent
	FY 2012: 210,000 kl
	FY 2013: 260,000 kl
	FY 2014: 320,000 kl
	FY 2015: 380,000 kl
	FY 2016: 440,000 kl
	FY 2017: 500,000 kl

[a]Ministry of Economy, Trade, and Industry (2010)

changes in land use, imports, and environmental conditions as well as domestic supply capacity and national mandates. Based on the analysis, we end the discussion with policy implications of moving toward sustainable biofuel.

15.2 Methods and Materials

Available future scenarios were reviewed for transportation usage of bioethanol and biodiesel. The national targets for bioethanol (Table 15.2) were set on the basis of Public Notice No. 242 issued by the Ministry of Economy, Trade, and Industry (METI) in 2010. The biodiesel targets in Table 15.2 followed the targets set by the MOE in 2006, but we modified them by shifting 5 years ahead from the original targets (i.e., interpreting the 2030 MOE target as the 2035 target for this analysis) because the actual diffusion of biodiesel has been delayed.

For analyzing each scenario, five options were prepared by considering the type of biomass, producer country, associated land use changes, competition with respect to food production, supply pattern, and transportation (Figs. 15.1 and 15.2).

We used three assessment indicators: carbon footprint (CF), WF, and EF. CFs and WFs for biofuel derived from different crops were collected extensively and

Table 15.2 Biofuel diffusion scenario for this study (Crude oil equivalent)

		2015	2025	2035
Assumption of fuel demand for transportation		Current demand (86,000,000 kl) gasoline (53,400,000 kl) diesel (32,600,000 kl)	80 % of the current demand gasoline (42,720,000 kl) diesel (26,080,000 kl)	50% of current demand gasoline (26,700,000 kl) diesel (16,300,000 kl)
Bioethanol	Assumption of bioethanol usage	Based on the target set by the Ministry of Economy, Trade, and Industry in 2010, the amount of bioethanol usage is assumed to increase from 210,000 kl crude oil eq. in 2011 by 60,000 kl crude oil eq. per year by promoting E3- and ETBT-added gasoline	The amount of bioethanol usage is assumed to continuously increase by 60,000 kl crude oil eq. every year from 210,000 kl crude oil eq. in 2011 by promoting E3-, E10-, and ETBT-added gasoline. (380,000 kl + 60,000 kl/year × 10 years = 980,000 kl)	The amount of bioethanol usage is assumed to continuously increase by 60,000 kl eq. crude oil every year from 210,000 kl crude oil eq. in 2011 by promoting E10- and ETBT-added gasoline. (980,000 kl + 60,000 kl/year × 10 years = 1,580,000 kl)
	Amount of bioethanol usage[a, b]	656,000 kl (380,000 kl)	1,692,000 kl (980,000 kl)	2,850,000 kl (1,580,000 kl)
	Domestic production	50,000 kl (30,000 kl)	950,000 kl (550,000 kl)	Ensure 2,850,000 kl not only by domestic bioethanol but also by imports from Brazil and Asian countries
	Import	606,000 kl (350,000 kl)	742,000 kl (430,000 kl)	

(continued)

Table 15.2 (continued)

		2015	2025	2035
Biodiesel	Assumption of biodiesel	High concentration of mixed biodiesel such as BDF 100 % and B20 will be promoted, whereas low-concentration mixed BDF such as B5 will be introduced extensively	Approximately one-third of the total demand for light diesel oil will be supplied by BDF, eco-diesel, and BTL (biomass–liquid). Domestic vegetable oil will be used to produce BDF and eco-diesel. Domestic wastes and forest biomass will be used for producing BTL	The total demand for light diesel oil will be supplied by BDF, eco-diesel, and BTL. Fulfill the demand by maximizing the utilization of domestic biomass resources and imports from Asian countries
	Amount of biodiesel[b]	11,000–16,000 kl (10,000–15,000 kl)	1,000,000 kl (900,000 kl)	2,000,000 kl (1,800,000 kl)
	Domestic production	11,000–16,000 kl (10,000–15,000 kl)	1,000,000 kl (900,000 kl) including the import from Asian countries, etc.	2,000,000 kl (1,800,000 kl) including import from Asian countries, etc.
	Import	Depending on the expansion of domestic production capacity		

[a]Ministry of Economy, Trade, and Industry (2010)
[b]Global Environmental Bureau, Ministry of Environment (2006): Report by the Promotion Council for Eco-Fuel Utilization, http://www.hkd.meti.go.jp/hokne/sui2nd_result/data1_7.pdf

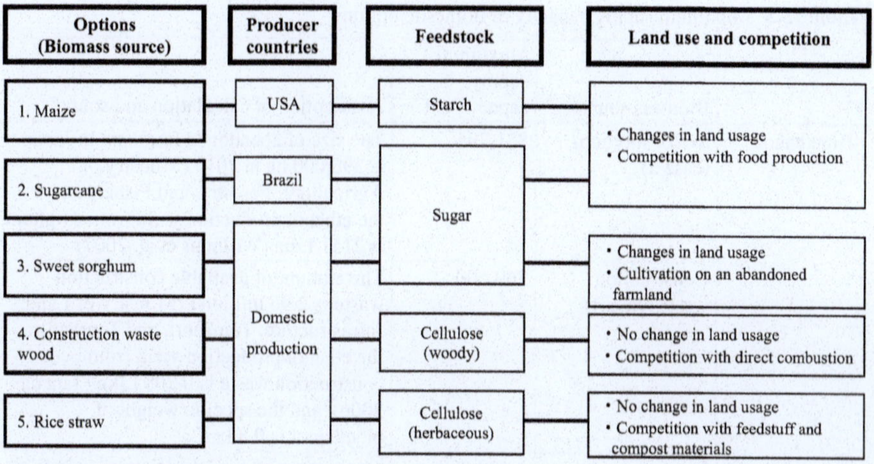

Fig. 15.1 Supply options for bioethanol in Japan

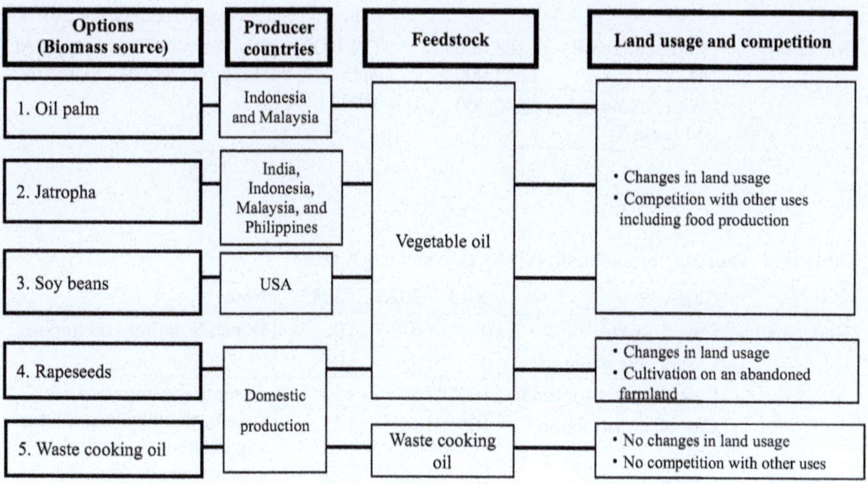

Fig. 15.2 Supply options for biodiesel in Japan

reviewed to identify differences among biomass sources. The maximum supply capacities of domestic options such as rice straw ethanol and waste cooking oil were calculated on the basis of domestic production and consumption of each biomass source (Table 15.3). Due to the variation in CF and WF values within the same biomass source, we used both upper-end and lower-end values as best case and worst case while calculating EF. Table 15.4 summarizes the domestic biofuel ratio (%) of each case and the target year. Unless Japan cannot expand the maximum supply capacity of the domestic options (Table 15.3), the domestic biofuel ratio will decrease owing to the increase in imported biofuel, which is necessary to fill the gap between domestic production and the targets, as described in Table 15.2.

Table 15.3 Maximum supply capacity of domestic options

	Biomass source	Maximum supply capacity (kL)	Assumption of calculation and source
Bioethanol	Sweet sorghum (case 3)	851,796	The size of abandoned farmland in Japan is 396,000 ha in 2010 (Ministry of Agriculture, Forestry, and Fishery), and the ethanol production yield from sorghum is 2151 L/ha (Williams et al. 2007)
	Construction waste (case 4)	769,600	The amount of available construction waste is 2.96 million t (Ministry of Land, Infrastructure, Transport, and Tourism), the ethanol production yield from construction waste is 0.208 L/kg (Taneda 2006), and the specific weight of bioethanol is 0.8 kg/L
	Rice straw (case 5)	1,600,080	The amount of available rice straw is 6.78 million t (METI 2007); the ethanol production yield from rice straw is 0.236 L/kg (National Institute of Advanced Industrial Science and Technology 2010)
Biodiesel	Rapeseed (case 4)	283,000–343,000	The BDF supply potentials from rapeseed and waste cooking oil were calculated by METI (2007)
	Waste cooking oil (case 5)	500,000	

Table 15.4 Domestic biofuel ratio (%) by case and target year

	Case	2015	2025	2035	Note
Bioethanol	Case 1: maze	0	0	0	Depends entirely on imports
	Case 2: sugarcane	0	0	0	
	Case 3: sweet sorghum	100	50	30	Assume the imported sugarcane ethanol to fill the gap between domestic production and targets
	Case 4: construction waste	100	46	27	
	Case 5: rice straw	100	95	56	
	Case 6: combination of domestically produced bioethanol	100	100	100	Depends entirely on domestically produced bioethanol
Biodiesel	Case 1: palm oil	0	0	0	Depend entirely on imports
	Case 2: *Jatropha*	0	0	0	
	Case 3: soybean	0	0	0	
	Case 4: rapeseed	100	28–34	14–17	Assume the imported palm oil biodiesel to fill the gap between domestic production and targets
	Case 5: waste cooking oil	100	50	25	
	Case 6: combination of domestically produced biodiesel	100	78–84	39–42	

15.2.1 Carbon Footprint

CF or $LCCO_2$ is one of the most popular indicators used in many LCA studies. CF can be defined as the total GHG emission due to biomass cultivation, extraction, transportation, the process of conversion to biofuel, and shipping of the biofuel. Today, CF is applied to the product labeling scheme in many countries.

15.2.2 Water Footprint

Water is needed for several processes in biofuel production. WF can be defined as the total annual volume of fresh water used to produce goods and services for consumption. WF consists of three components: the green WF, blue WF, and gray WF (Worldwatch Institute 2007). The green WF refers to rainwater that evaporates during production, mainly during crop growth. The blue WF is the surface- and groundwater used for irrigation that evaporates during crop growth. The gray WF is the amount of water needed to dilute pollutants discharged into the natural water system to the extent that the quality of the ambient water remains above agreed-upon water quality standards.

15.2.3 Ecological Footprint

EF is a tool to measure human demand by comparing with Earth's ecological capacity to regenerate. It indicates the amount of biologically productive land and sea area needed to regenerate the resources consumed by a human population and to absorb its wastes (Rees 1992; Wackernagel 1994). Conceived in 1990 by Mathis Wackernagel and William Rees at the University of British Columbia, EF has been widely used by scientists, businesses, governments, agencies, individuals, and institutions to monitor ecological resource use and assess our pressure on Earth's system. The following equation was used to calculate EF in this study. Wackernagel and Rees (1995) selected 6.6 mt as their average value for the total CO_2 sequestered by the world's forests. Therefore, we also used the value of 6.6 Mg/ha for CO_2 sequestration. This value would be 3.2 Mg/ha (Greenhouse Gas Inventory Office of Japan 2010) by assuming the offset CO_2 emissions from the forests in Japan:

$EF(ha) = EF_{cf} + EF_{harvest} + EF_{water}$ where

EF_{cf} = Forest cover (ha) needed to assimilate CO_2 emissions from the biofuel supply (i.e., CF)

$EF_{harvest}$ = Farmland cover (ha) needed to harvest crops or vegetables for biofuel

EF_{water} = Water catchment area (ha) needed to collect the total water volume required to grow biofuel crops and vegetables (the blue *WF* and the green *WF*)

15.3 Results

15.3.1 CF, WF, and EF per Unit Amount

15.3.1.1 Carbon Footprint

Table 15.5 and Fig. 15.3 summarize the net life-cycle GHG emissions from biofuels derived from different biomass sources. Within the same type of biofuel such as corn ethanol, different studies report different values depending on the researcher, production system, and accounting boundary. Until 2005, most of the studies on corn ethanol showed a corn ethanol CF slightly larger than that of gasoline, but studies after 2006 have demonstrated a 20 % or even greater GHG reduction by

Table 15.5 Life-cycle GHG emissions excluding those due to changes in land usage

		Year	Net GHG emissions (g-CO2/MJ)	Notes	Source
Reference	Gasoline		94.0		[a]
	Gasoline		92.0	[b]	
	Gasoline (Japan)		81.7	[c, d]	
	Diesel		82.3	[e]	
Corn ethanol	Marland and Turhollow	1991			[f]
	Lorenz and Morris	1995			[f]
	Wang	2001	71.0		[a]
	Graboski	2002	99.0		[a]
	Shapouri et al.	2002			[f]
	Patzek	2004	121.0		[a]
	Shapouri et al.	2004	61.0		[a]
	Pimentel et al.	2005	116.0		[a]
	de Oliveira et al.	2005	98.0		[a]
	Kim and Dale	2005			[f]
	Farrell et al.	2006	87.0		[a]
	Hill et al.	2006	84.9		[e]
	Fargione et al.	2008	78.3		[g]
	Serchinger et al.	2008	74.0		[b]
	Toyota Motor Corporation and Mizuho Information and Research Institute	2008	81.4	Maximum case	[h]
			54.0	Minimum case	[h]
	EU directive 2009/28/EC	2009	43.0	Community produced (natural gas as process fuel in CHP plant)	[i]

(continued)

Table 15.5 (continued)

		Year	Net GHG emissions (g-CO2/MJ)	Notes	Source
Sugarcane ethanol	Fargione et al.	2008	17.9		g
	Toyota Motor Corporation and Mizuho Information and Research Institute	2008	14.8	Maximum case	h
			14.5	Minimum case	h
	EU directive 2009/28/EC	2009	24.0		i
	Ministry of Economy, Trade, and Industry, Japan	2010	32.7	Including shipping from Brazil to Japan (13.9 g-CO2eq/MJ)	c, d
Sugar beet ethanol	EU directive 2009/28/EC	2009	40.0		i
Sweet sorghum ethanol	Xunmin et al.	2009	36.3	China	j
Wheat ethanol	EU directive 2009/28/EC	2009	70.0	Process fuel not specified	i
			44.0	Natural gas process fuel in CHP plant	i
			26.0	Straw gas process fuel in CHP plant	i
Soybean biodiesel	Hill et al.	2006	49.0		e
	EU directive 2009/28/EC	2009	58.0		i
	Xunmin et al.	2009	41.9	China	j
Palm biodiesel	Fargione et al.	2008	37.0		g
	Toyota Motor Corporation and Mizuho Information and Research Institute	2008	13.4		h
	Yee et al.	2009	31.7		k
	EU directive 2009/28/EC	2009	68.0	Process not specified	i
			37.0	Process with methane capture at oil mill	i
Rapeseed biodiesel	EU directive 2009/28/EC	2009	52.0		i
Jatropha biodiesel	Prueksakorn and Gheewala	2005	16.5		l
	Tobin and Fulford	2006	56.7		m
	Xunmin et al.	2009	34.6	China	j

(continued)

Table 15.5 (continued)

		Year	Net GHG emissions (g-CO2/MJ)	Notes	Source
Cellulosic bioethanol	Farrell et al.	2006	11.0		a
	Serchinger et al.	2008	27.0	Switch grass	b
	Toyota Motor Corporation and Mizuho Information and Research Institute	2008	50.3	USA (cellulosic) maximum case	h
			25.2	USA (cellulosic) minimum case	h
			20.3	Forest thinning's (Japan) maximum case	h
			7.9	Forest thinning's (Japan) minimum case	h
	EU directive 2009/28/EC	2009	13.0	Wheat straw ethanol	i
			22.0	Waste wood ethanol	i
			25.0	Farmed wood ethanol	i

[a]Farrell et al. (2006)
[b]Searchinger et al. (2008)
[c]Agency for Natural Resources and Energy, Ministry of Economy, Trade, and Industry (2010)
[d]Ministry of Economy, Trade, and Industry (2010)
[e]Hill et al. (2006)
[f]Hammerschlag (2006)
[g]Fargione et al.(2008)
[h]Toyota Motor Corporation and Mizuho Information and Research Institute (2008)
[i]Directive 2009/28/EC of the European Parliament and of the Council of April 23, 2009, on the promotion of the use of energy from renewable sources and amending and subsequently repealing Directives 2001/77/EC and 2003/30/EC
[j]Xunmin et al. (2009)
[k]Kian et al. (2009)
[l]Tobin and Fulford (2005)
[m]Prueksakorn and Gheewala (2006)

gasoline. Sugarcane ethanol has a smaller CF than that of corn ethanol, which is equivalent to one-fifth of the gasoline GHG emission. This relative advantage of sugarcane is because the bagasse—a by-product of the sugarcane plant—can be used as an energy source in ethanol refinery. METI's Public Notice No.242 (2010) specifies that CF from bioethanol should be less than 50 % of that from gasoline (81.7 g-CO_2eq/MJ).

CF from soybean biodiesel is reported to be approximately half that of conventional diesel. CF from palm oil biodiesel is even smaller than that of soybean biodiesel if we ignore the methane emissions from the conversion of peatland to oil palm plantations, a common occurrence in Indonesia and Malaysia.

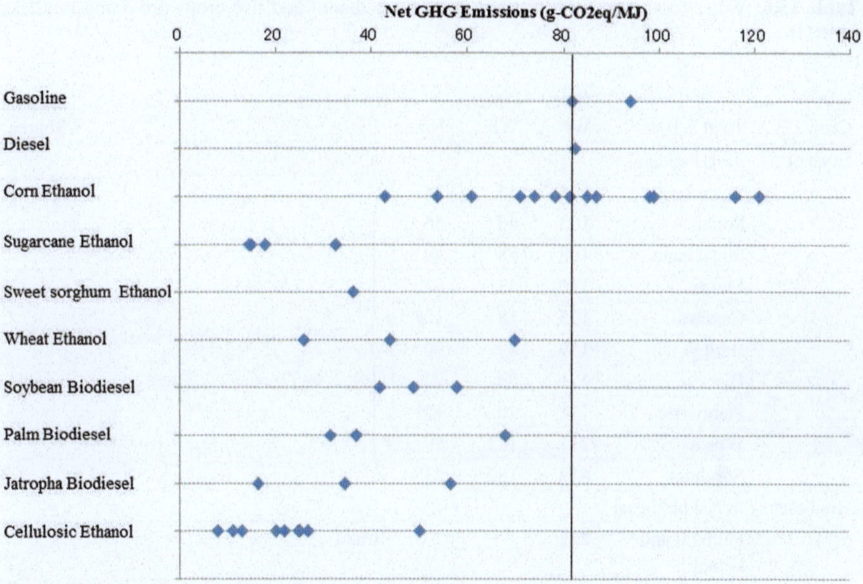

Fig. 15.3 Life-cycle GHG emissions (carbon footprint) of various biofuels

15.3.1.2 Water Footprint

Table 15.5 summarize WF per unit amount of fuel. Gerbens–Leenes et al. (2009a) report that WF of biodiesel is generally greater than that of bioethanol while using global averages. The global average WF of biodiesel crops ranges from 394 to 574 m^3/GJ biodiesel. *Jatropha* is famous for being tolerant to wasteland, but its requirement for water is greater than many other energy crops, which implies that water availability may be one of the constraints for *Jatropha* biodiesel supply.

The global average WF of bioethanol crops ranges from 59 to 419 m^3/GJ. WFs of sugar beet, potato, and sugarcane are 59, 103, and 108 m^3/GJ, respectively, whereas sorghum (419 m^3/GJ) has the largest WF of all ethanol crops (Table 15.6).

These results suggest that switching to biomass energy may result in an increased demand for fresh water, which eventually will intensify the competition between water usage for food production and energy (Bazilian et al. 2011).

15.3.1.3 Ecological Footprint per Unit Amount of Biofuel

EFs per unit of biofuel are compared according to cases in Fig.15.4. Producing bioethanol from sorghum and maize results in a larger EF than production from other biomass sources. Using construction waste wood is the best option for minimizing EF (Fig. 15.4a). Biodiesel from *Jatropha* and soybean yields an EF two to

Table 15.6 Water footprints for ten crops providing ethanol and five crops providing biodiesel (m^3/GJ)

Crop	Total WF	Blue WF	Green WF	Note		Source
Ethanol	m^3/GJ ethanol					
	Sugar beet	59	35	24	Total weighted global average	[a]
	Potato	103	46	56		[a]
	Sugar cane	108	58	49		[a]
	Maize	110	43	67		[a]
	Cassava	125	18	107		[a]
	Barley	159	89	70		[a]
	Rye	171	79	92		[a]
	Paddy rice	191	70	121		[a]
	Wheat	211	123	89		[a]
	Sorghum	419	182	238		[a]
Biodiesel	m^3/GJ biodiesel					
	Palm oil and kernel	247			Brazil	[b]
	Sunflower	377			Average of the Netherlands, the USA, Brazil, and Zimbabwe	[b]
	Soybean	394	217	177	Total weighted global average	[a]
	Rapeseed	409	245	165		[a]
	Jatropha	574	335	239		[a]

[a]Gerbens–Leenes et al. (2009a)
[b]Gerbens–Leenes et al. (2009b)

three times greater than other cases, and converting waste cooking oil to BDF is the best among all cases (Fig. 15.4b). Palm oil shows the smallest EF among three cases of imported biodiesel from other countries.

15.3.2 Scenario Analysis

Considering the targets for 2015, 2025, and 2035, different cases to achieve the targets (Figs. 15.1 and 15.2), the maximum supply capacity of each domestic biomass source (Table 15.3), and the domestic biofuel ratio (Table 15.4), we calculated CF, WF, and EF from 2015 to 2023 (Figs. 15.5, 15.6, 15.7, and 15.8). In addition to the five cases for each biofuel described in Figs. 15.1 and 15.2, we prepared a sixth case that maximizes the domestic biomass sources by combining sorghum, construction waste wood, and rice straw for bioethanol and by combining rapeseed and waste cooking oil for biodiesel (Table 15.4).

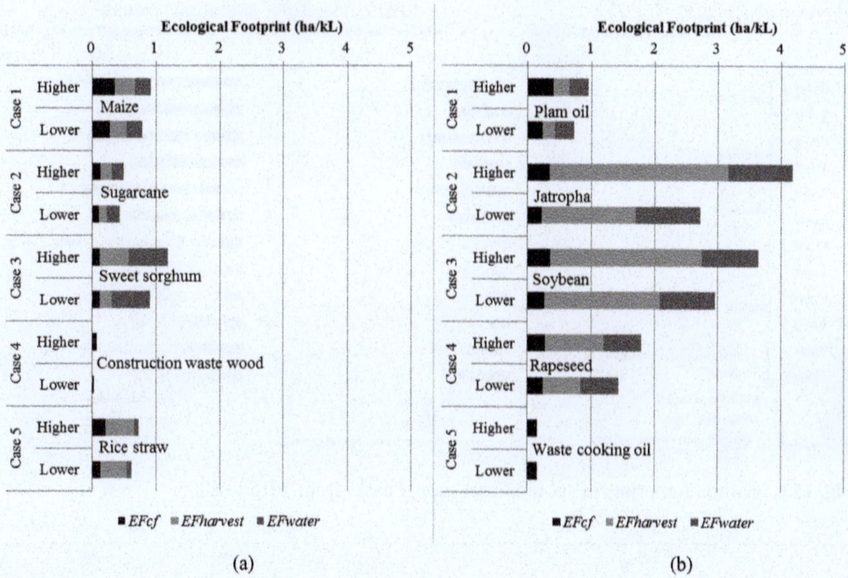

Fig. 15.4 Ecological footprint per unit of biofuel for five cases each of (**a**) bioethanol and (**b**) biodiesel

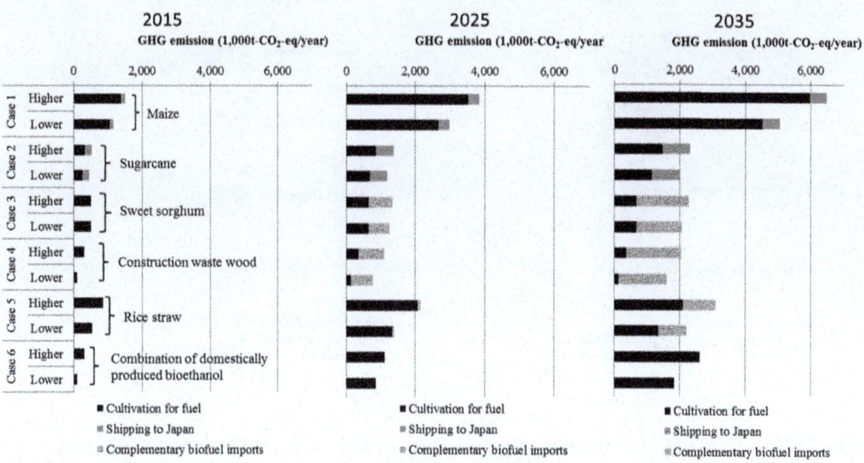

Fig. 15.5 Carbon footprints of six bioethanol supply cases from 2015 to 2035

In terms of GHG emissions (CF), imported maize bioethanol shows the worst performance of the six cases, whereas bioethanol from sweet sorghum and construction waste wood shows better performances (Fig. 15.5). Bioethanol from rice straw emits more GHGs than other domestic cases (cases 3, 4, and 6). The difference between sugarcane ethanol imported from Brazil (case 2) and ethanol from

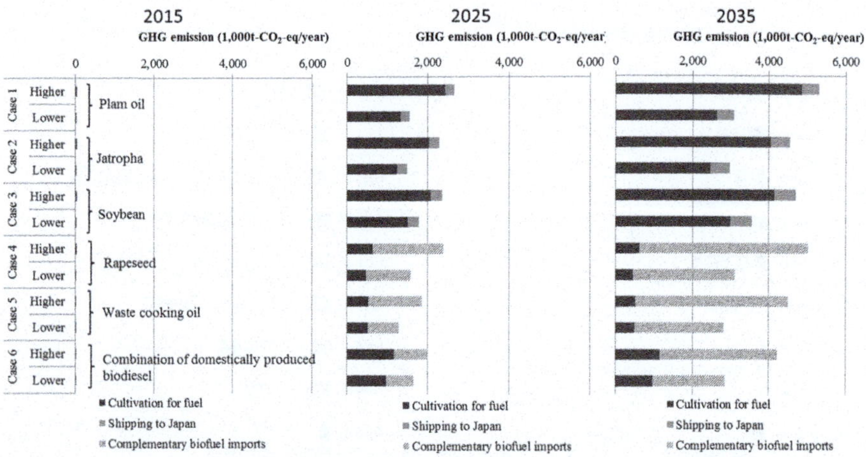

Fig. 15.6 Carbon footprints of six biodiesel supply cases from 2015 to 2035

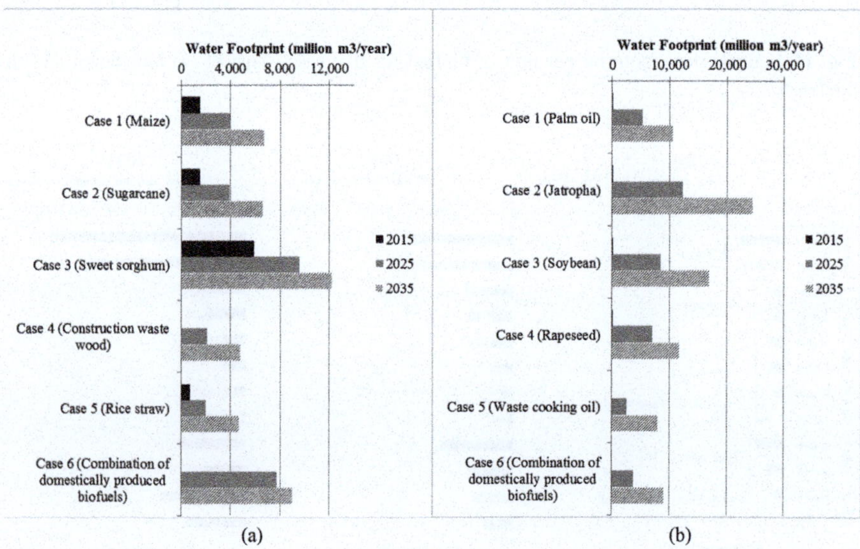

Fig. 15.7 Water footprints of six supply cases from 2015 to 2035. (**a**) Bioethanol (**b**) Biodiesel

domestic construction waste wood (case 4) is reduced in 2035 because imports of complementary bioethanol are increased to achieve the target.

GHG emissions from the domestic biodiesel cases (cases 4–6) tend to be lower than the importing cases, but the differences are not as significant as those in the bioethanol cases (Fig. 15.6). The combination of all domestic BDFs (case 6) gives the best result of all the cases.

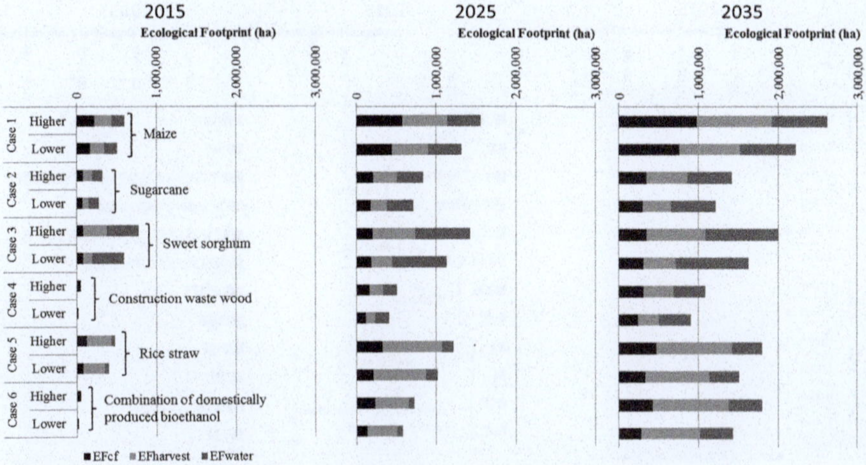

Fig. 15.8 Ecological footprints of six bioethanol supply cases from 2015 to 2035

Among the bioethanol WFs from the six cases, sweet sorghum (case 3) shows the largest WF (Fig. 15.7a). Therefore, case 6, which maximizes domestic biodiesel, indicates a larger WF than that of construction waste wood (case 4) and rice straw (case 5). *Jatropha* (case 2) requires the maximum amount of water out of any of the other cases investigated in this study (Fig. 15.7b). Palm oil (case 1) and domestic rapeseed (case 4) show similar WF performances. Waste cooking oil (case 5) is the best option in terms of WF, even considering the complementary import of biodiesel (palm oil) to fill the gap between the maximum supply capacity of waste cooking oil and the national target.

Figure 15.8 summarizes EFs of all bioethanol cases from 2015 to 2035. Construction waste wood shows the smallest EF out of all the cases, whereas maize ethanol is calculated to have the largest EF. In 2035, maximizing the domestic sources (case 6) would not be the best option because the performance of bioethanol is almost similar to that of sugarcane (case 2) and rice straw (case 5), which suggests that care should be taken while selecting combinations of available options to minimize EF in longer term.

Jatropha has the largest EF of all the cases, with soybean coming in the second place (Fig. 15.9) because of the large land area required to harvest it ($EF_{harvest}$) and the catchment area required for water (EF_{water}). EF of waste cooking oil (case 5) was the smallest of all the cases, but the EFs of palm oil (case 1), rapeseed (case 4), and the combination of domestically produced biodiesel (case 6) were all less than 2 million ha. The results demonstrate that importing biodiesel produced from *Jatropha* and soybean does not make sense in terms of EF because their EFs are three to four times larger than those of other cases.

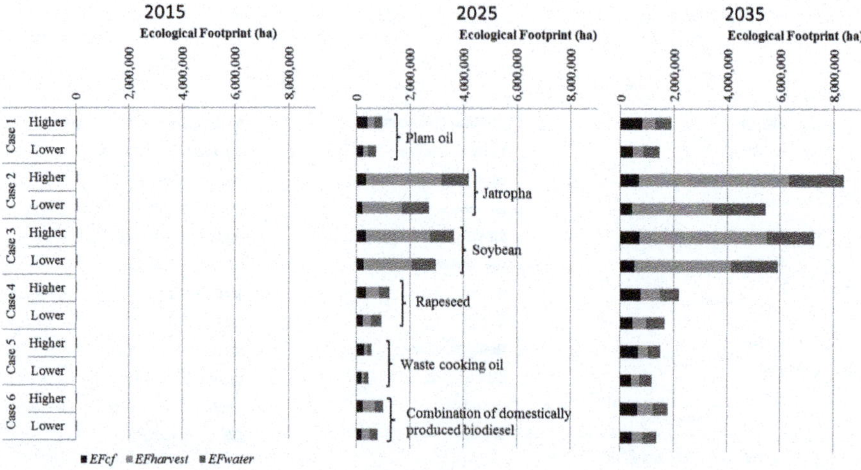

Fig. 15.9 Ecological footprints of six biodiesel supply cases from 2015 to 2035

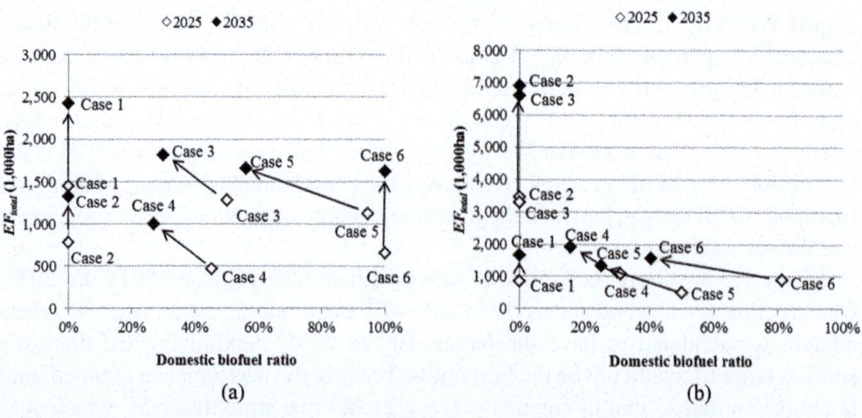

Fig. 15.10 EF_{total} and domestic biofuel ratio by case. (**a**) Bioethanol (**b**) Biodiesel

15.4 Discussion and Conclusion

An integrated sustainability assessment model of biofuel that uses several biomass sources was developed in this chapter. Figure 15.10 summarizes the results of the scenario analysis, which uses six different cases to achieve Japan's national target for bioethanol and biodiesel. This figure suggests that Japan needs to import more than 40 % of its bioethanol to achieve the national target in 2035, except in case 6 (maximizing domestically produced bioethanol) (Fig. 15.10a). Similarly, Japan needs to import at least 59 % of its total biodiesel to achieve the 2035 target

Fig. 15.11 Relationships between EF_{cf} and $EF_{harvest}$ in six bioethanol cases

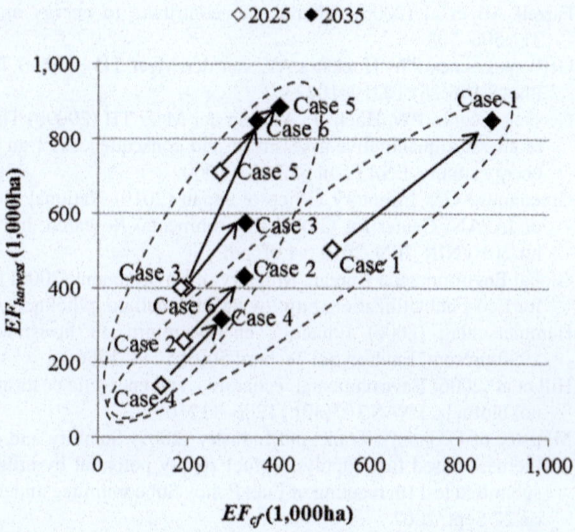

(Fig. 15.10b). In general, a dependency on the imported biofuel or a self-sufficiency in biofuel production has an influence on the level of EF_{total}.

This assessment model can provide not only the overall ecological footprint for each case but also a detailed breakdown of EF_{cf}, $EF_{harvest}$, and EF_{water}. This allows us to identify relationships across these indicators. For example, Fig. 15.11 indicates the linkage between EF_{cf} and $EF_{harvest}$ in six bioethanol cases, which suggests that EF_{cf} in general increases $EF_{harvest}$, but we can find different paths (regression lines) with steeper slopes, such as case 6, and those with moderate slopes, such as cases 1, 2, and 4. This means that the same reduction in GHG emission results in different levels of $EF_{harvest}$ depending on the case chosen by the government. It is highly recommended that the government applies multi-criteria sustainability assessment as demonstrated by this chapter in addition to conventional cost-benefit analysis prior to making a policy decision to expand biofuel production and import.

References

Agency for Natural Resources and Energy, Ministry of Economy, Trade and Industry (2010) Commission report on sustainability criteria for introducing biofuel (Written in Japanese)

Bazilian M, Rogner H, Howells M, Hermann S, Arent D, Gielen D, Steduto P, Mueller A, Komor P, Tol RSJ, Yumkella KK (2011) Considering the energy, water and food nexus: towards an integrated modelling approach. Energy Policy 39:7896–7906

DIRECTIVE 2009/28/EC OF THE EUROPEAN PARLIAMENT AND OF THE COUNCIL of 23 April 2009 on the promotion of the use of energy from renewable sources and amending and subsequently repealing Directives 2001/77/EC and 2003/30/EC

Fargione J et al (2008) Land clearing and the biofuel carbon debt. Science 319:1235–1238

Farrell AE et al (2006) Ethanol can contribute to energy and environmental goals. Science 311:506–508

Gerbens-Leenes PW, Hoekstra AY, van der Meer TH (2009a) The water footprint of bioenergy. PNAS 106(25):10219–10223

Gerbens-Leenes PW, Hoekstra AY, van der Meer TH (2009b) The water footprint of energy from biomass: a quantitative assessment and consequences of an increasing share of bioenergy in energy supply. Ecol Econ 68:1052–1060

Greenhouse Gas Inventory Office of Japan (2010) National Greenhouse Gas Inventory Report of JAPAN. Center for Global Environmental Research. http://www-gio.nies.go.jp/aboutghg/nir/2010/NIR_JPN_2010_v4.0E.pdf

Global Environmental Bureau, Ministry of Environment (2006) Report by the Promotion Council for Eco-Fuel Utilization. http://www.hkd.meti.go.jp/hokne/sui2nd_result/data1_7.pdf

Hammerschlag (2006) Ethanol's energy return on investment: a survey of the literature 1990-present. Environ Sci Technol 50(6):1744–1750

Hill et al (2006) Environmental, economic, and energetic costs and benefits of biodiesel and ethanol biofuels. PNAS 103(30):11206–11210

Ministry of Economy, Trade and Industry (2007) Stability and economic efficiency of domestic biomass-based fuel supply: biofuel supply potential by utilizing domestic natural resources, submitted to 11th meeting of Fuel Policy Subcommittee under Advisory Committee for Energy on 22 Sept. 2007

Ministry of Economy, Trade and Industry (2010) Public Notice No. 242, judgment criteria for the use of non-fossil energy sources by oil refiners (Written in Japanese)

National Institute of Advanced Industrial Science and Technology (2010) Environmental impact assessment report on ethanol production from rice straw. http://lca-forum.org/research/lime/pdf/04_a.pdf

Prueksakorn K, Gheewala SH (2006) Energy and greenhouse gas implications of biodiesel production from *Jatropha curcas* L., Proceedings of The 2nd Joint International Conference on "Sustainable Energy and Environments (SEE 2006)": Anonymous., Bangkok, Thailand, Nov. 21–23, 2006

Rees WE (1992) Ecological footprints and appropriated carrying capacity: what urban economics leaves out. Environ Urban 4(2):121–130

Searchinger T et al (2008) Use of U.S. croplands for biofuels increases greenhouse gases through emissions from land-use change. Science 319:1238–1240

Taneda D (2006) Concentrated sulfuric acid biomass ethanol process. Cellulose Commun 13(2):49–52

Tobin J, Fulford DJ (2005) Life cycle assessment of the production of biodiesel from Jatropha. Msc dissertation. The University of reading, 2005

Toyota Motor Corporation & Mizuho Information and Research Institute (2008) Well-to-wheel assessment of transportation fuel: study report on well-to-tank GHG emission from transportation fuel production including biofuel (Written in Japanese)

Wackernagel M (1994) Ecological footprint and appropriated carrying capacity: a tool for planning toward sustainability (PhD thesis). Vancouver, Canada: School of Community and Regional Planning. The University of British Columbia. OCLC 41839429

Wackernagel M, Rees W (1995) Our ecological footprint: reducing human Impact on the earth. New Society Publishers, Gabriola Island/Philadelphia

Williams RB, Jenkins BM, Gildart MC (2007) California biofuel goals and production potential. 15th European biomass conference and exhibition, 7–11 May, 207, Berlin, Germany

Worldwatch Institute (2007) Biofuels for transport. Global potential and implications for sustainable energy and agriculture. Earthscan, London

Xunmin O, Xiliang Z, Shiyan C, Qingfang G (2009) Energy consumption and GHG emissions of six biofuel pathways by LCA in China. Appl Energy, article in press

Yee KF, Tan KT, Abdullah AZ, Lee KT (2009) Life cycle assessment of palm biodiesel: revealing facts and benefits for sustainability. Appl Energy, (in press)

Correction to: Stakeholder Perceptions of the Ecosystem Services and Human Well-Being Impacts of Palm Oil Biofuels in Indonesia and Malaysia

Raquel Moreno-Peñaranda, Alexandros Gasparatos, Per Stromberg, Aki Suwa, and Jose A. Puppim de Oliveira

Correction to:
Chapter 10 in: K. Takeuchi et al. (eds.), *Biofuels and Sustainability*, Science for Sustainable Societies, https://doi.org/10.1007/978-4-431-54895-9_10

The chapter was originally published with an incorrect citation and its reference. The occurrence of reference and its citation has been removed from page numbers 170 (line 26 & 27) and 146 (line 14).

The updated version of this chapter can be found at
https://doi.org/10.1007/978-4-431-54895-9_10

K. Takeuchi et al. (eds.), *Biofuels and Sustainability*, Science for Sustainable Societies, https://doi.org/10.1007/978-4-431-54895-9_16

The manufacturer's authorised representative in the EU is Springer Nature Customer Service Centre GmbH, Europaplatz 3, 69115 Heidelberg, Germany. If you have any concerns regarding our products, please contact ProductSafety@springernature.com

Printed and bound by CPI Group (UK) Ltd, Croydon, CR0 4YY
30/09/2026
02225454-0004